PROJECT MANAGEMENT FOR ENGINEERS

PROJECT MANAGEMENT FOR ENGINEERS

J. Michael Bennett • Danny S. K. Ho

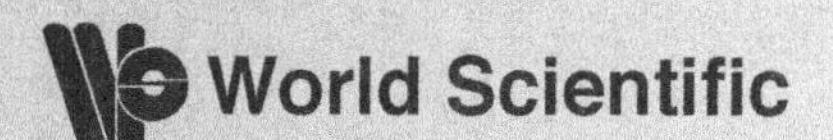

NEW JERSEY • LONDON • SINGAPORE • BEIJING • SHANGHAI • HONG KONG • TAIPEI • CHENNAI

Published by

World Scientific Publishing Co. Pte. Ltd.
5 Toh Tuck Link, Singapore 596224
USA office: 27 Warren Street, Suite 401-402, Hackensack, NJ 07601
UK office: 57 Shelton Street, Covent Garden, London WC2H 9HE

Library of Congress Cataloging-in-Publication Data
Bennett, J. Michael.
Project management for engineers / J. Michael Bennett, University of Ontario, Canada, Danny Ho, NFA Estimation Inc., Canada.
pages cm
Includes bibliographical references and index.
ISBN 978-9814447928 (hardcover : alk. paper)
1. Project management. I. Ho, Danny. II. Title.
T56.8.B458 2013
658.4'0402462--dc23

2013023791

British Library Cataloguing-in-Publication Data
A catalogue record for this book is available from the British Library.

Typeset by Stallion Press
Email: enquiries@stallionpress.com

Printed in Singapore

Endorsement

"After many years of managing Software Projects for IBM Software Group, this book is a reminder of the discipline and need to follow Best Practices in Project Management . . . if corporations expect a predictable and successful product delivery.

The material is concise, well written, and easy to understand, with good examples, clarifying situations or issues. The effective use of PMBOK helps an aspiring PM (Project Manager) with his/her certification, and provides an experienced PM with a good reference guide.

With a growing global appetite for software applications to support both business and consumer demand, there is a definite need for a disciplined approach to Project Management in Software Engineering, when we are expected to deliver on time and within budget.

Highly recommended . . . a persuasive case for doing business in the digital (software) age, when trial & error are no longer acceptable."

James van der Voet,
Senior Vice President, Technology (retired),
Allstream/ADS

"An excellent and practical book that every student in engineering should read and therefore learn the easy way. The alternative is to learn the hard way... on the job."

René Tinawi, Ph.D., eng.,
Professor Emeritus,
École Polytechnique de Montréal

"Project Management remains both a well-recognized and ill-recognized profession. Despite its undeniable value for those who understand the profession, I frequently encounter professionals — often in positions of considerable responsibility — who lack even an elementary understanding of the considerable value that project management expertise and competencies bring not only to projects, but to management teams. For this to change, we need to take the long view, and education is essential. The introduction of project management principles and practices — and the profession itself — to university undergraduates is essential, whether or not they eventually choose to engage in the profession.

Michael Bennett and Danny Ho are to be commended for doing just this: writing a book on project management that connects to engineering students, and perhaps to a much larger audience. The book is easy to read, and immediately piques the reader's interest as we follow one of the authors in his personal journey through the computer revolution of the second half of the 20th century. This bit of history provides fascinating insight on evolution of computer and information technology, educates the reader on the need for greater sophistication as projects grow in scope and complexity, and leverages key professional milestones as pretexts for introducing fundamental concepts.

However, those who would believe this book to be no more than an introduction will be profoundly mistaken: the authors' knowledge of project management breathes real life into the many topics that make up the Project Management Body of Knowledge through many examples which, in turn, create bridges to deeper understanding. The readers — students or laypersons — will find much here to stimulate their interest in project management. This book may well provide the foundation for far greater appreciation and respect for the profession over time."

Richard J. Marceau, PEng, FCAE,
Vice President (Research),
Memorial University of Newfoundland
President,
The Canadian Academy of Engineering (2012–2014)

Preface

There are a lot of books, publications, and experience reports on project management from many qualified authors. Why is there a need for yet another book on the same subject matter? The authors of this book, Mike and Danny, believe that this book blends together many years of professional and industrial practices, with consulting and teaching experience at the professional training, graduate, and senior undergraduate levels. The book is suited for students who obtain their first exposure to project management with simple practical examples (such as a forward and backward path calculation). It is also suited for professionals who need more practical experience in project management (e.g., in handling outsourcing/offshoring and downsizing/right-sizing issues). Indeed, we stress the importance of integrating engineering technical issues with business issues, wrapped in a project management blanket. There is also coverage on topics at the advanced (research) level (e.g., benefit-cost analysis of offshoring) for those interested in additional challenges.

Project management is a core competency for all engineering disciplines. Although the focus of this book is on engineering projects, the theory and practice presented are equally applicable to any industry/domain. A lot of the material in the book is based on the Project Management Body of Knowledge (PMBOK®). The discussion also includes other project management standards/processes (such as IEEE, PRINCE, etc.). The authors believe that this book is a good text book for undergraduate engineers and scientists as well as a good reference for professionals preparing to practice or practicing project management.

Acknowledgements

To Sandy: my inspiration before, now and always!

J. Michael Bennett

I did not realize that writing a book involved such a tedious and complicated process. It took years to organize all the material and put everything down in an orderly fashion. The writing all began in 2008 when I was confined to a nice guest room in a Cuba resort for almost one week while hurricane Ike made landfall. Overall, I still enjoyed the entire process of writing this book to the fullest. Thanks to World Scientific that made this publication possible.

First and foremost, I must extend my deepest and wholehearted appreciation to my parents Kam Min Ho (Mr.) and Sui Yin So-Ho (Mrs.) for their everlasting love, guidance, care and support throughout my entire life. I still enjoy every single minute of being their "home boy". Special thanks to my best friend and former colleague Mr. John Yat Ming Chan in providing much help and encouragement to my writing, especially in keeping the content of the book up-to-date as the years slipped by.

There are two very special individuals who influenced my studies and career the most. Dr. J. Michael Bennett, my teacher, thesis supervisor, career mentor and life-time reference, whom I gave the nickname "my Good Dean (GD)", is indeed the very best of the best. GD, thanks for allowing me to co-author and share this writing venture with you! Mr. James van der Voet, with knowledge in all aspects of technical and non-technical matters as vast as an endless ocean, is the expert and professional that I can always count on, draw upon and learn from each time I was given the opportunity to dialogue with, regardless of the duration (short and long).

There is always no replacement for the fundamentals, that is solid and sound education. I must mention the University of Western Ontario, London, Ontario, Canada that gave me the opportunity for undergraduate and graduate studies, to further my research as an Adjunct Research Professor with many top-notch researchers (amongst them Dr. Luiz F. Capretz) and to supervise many outstanding PhD and Masters students. Last, but not the least, I must also mention Saint Louis School, Hong Kong that fulfilled my highly rewarding and memorable primary and secondary education.

Danny Siu Kau Ho

About the Authors

Dr. J. Michael Bennett was the Associate Dean of Engineering at the University of Ontario Institute of Technology (UOIT). Prior to this, he was the Program Director of Electrical and Computer Engineering at the University of Western Ontario (UWO). He developed the Software Engineering Programs at UOIT and UWO, and led both programs to multiple CEAB certifications. He also worked with UWO and Cisco authorities to build the CCNA router laboratory at UWO. Before introducing Software Engineering to UWO, he was a professor at the Department of Computer Science. He has also taught at the University of Ottawa, the Institute for Government Informatics Professionals, Motorola University, and IMPAC University. His research areas include software engineering, software metrics, project management, computer communications, operating system, numerical analysis, etc. Throughout his academic career, he received over $2.34 million in research and industrial grants for over 20 projects. Dr. Bennett has supervised over 30 graduate students and over 40 undergraduate students, published many books, articles, and scientific papers. He also served as a contributor on 12 refereed ISO standards submissions. He is currently a member of the Professional Engineers Ontario (PEO) and a Project Management Professional (PMP). Contact email: Michael.Bennett@uoit.ca

Mr. Danny Siu Kau HO is an independent management consultant and adviser for two startup companies. Prior to this, he held senior management and technical positions at Motorola Canada Limited, Nortel Networks Corporation, and IBM Canada Limited. He has also been appointed as an Adjunct Research Professor at the Department of Software Engineering, Faculty of Engineering, the University of Western Ontario (UWO). Throughout his professional career, he has led programs in the areas of wire-line, RF and infrared development; eMarketing, mCommerce, desktop application deployment, reuse, and software development environment. He is the co-owner of two patents on neuro-fuzzy system for estimation His areas of special interest include software estimation, project management, object-oriented software development, and complexity analysis. Danny received his Honours Bachelor of Science in Computer Science with Electrical Engineering, and Master of Science in Computer Science from UWO. He is currently a member of the Professional Engineers Ontario (PEO) and a Project Management Professional (PMP). Contact email: danny@nfa-estimation.com

Contents

Acronyms

AC	Actual Cost
ANSI	American National Standards Institute
AOA	Activity-On-Arrow
AON	Activity-On-Node
BAC	Budget At Completion
BCR	Benefit-Cost Ratio
BNF	Backus-Naur Form
CC	Configuration Control
CCB	Change Control Board
CFO	Chief Financial Officer
CI	Configuration Item
CL	Center Line
CM	Configuration Management
CMM	Capability Maturity Model
CoQ	Cost of Quality
CPAF	Cost Plus Award Fee
CPFF	Cost Plus Fixed Fee
CPI	Cost Performance Index
CPIF	Cost Plus Incentive Fee
CR	Change Request
CR	Cost Reimbursement
CRB	Change Review Board

CRR	Comparative Risk Ranking
CSOW	Contract Statement Of Work
CV	Cost Variance
DDD	Detailed Design Document
DoE	Design of Experiment
EAC	Estimate At Completion
ECO	Engineering Change Order
EF	Early Finish
EMV	Expected Monetary Value
ES	Early Start
ETC	Estimate To Complete
EV	Earned Value
EVM	Earned Value Management
FFP	Firm Fixed Price
FIRO-B	Fundamental Interpersonal Relationship Orientation Behaviour
FMEA	Failure Mode and Effect Analysis
FP	Function Point
FP-EPA	Fixed Price — Economic Price Adjustment
FPIF	Fixed Price with Incentive Fee
FPPM	Function Point Per Month
FR	Functional Requirement
FV	Future Value
FW	Future Worth
ICC	Integrated Change Control
ICCP	Integrated Change Control Plan
IEEE	Institute of Electrical and Electronics Engineers
IFB	Invitation For Bid
ISO	International Organization for Standardization
IT	Information Technology

IV&V	Independent Verification and Validation
JAD	Joint Application Development
LCC	Life Cycle Costing
LCL	Lower Control Line (Limit)
LCM	Least Common Multiple
LF	Latest Finish
LOI	Letter Of Intent
LS	Latest Start
LSL	Lower Specification Limit
MARR	Minimum Attractive Rate of Return
MIPS	Million of Instructions Per Second
MOU	Memorandum Of Understanding
MTTD	Mean Time To Defect
MTTF	Mean Time To Failure
MTTR	Mean Time To Recovery
MTTR	Mean Time To Repair
NFR	Non-Functional Requirement
OOM	Order Of Magnitude
OPM3	Organizational Project Management Maturity Model
PDCA	Plan Do Check Act
PERT	Performance Evaluation and Review Technique
PM	Project Manager/Program Manager
PMBOK®	Project Management Body Of Knowledge
PMI	Project Management Institute
PMO	Project Management Office
PMP	Project Management Plan
PMP	Project Management Professional
PRINCE	PRoject IN Controlled Environment
PSS	Project Scope Statement
PV	Planned Value

PV	Present Value
PW	Present Worth
QAP	Quality Assurance Plan
QFD	Quality Function Deployment
QMP	Quality Management Plan
RACI	Responsible, Accountable, Consulted, Informed
RAM	Responsibility Assignment Matrix
RBS	Risk Breakdown Structure
RFI	Request For Information
RFP	Request For Proposal
RFQ	Request For Quotation
RFT	Request For Tender
RMP	Requirements Management Plan
ROA	Return On Asset
ROI	Return On Investment
ROR	Rate Of Return
ROS	Return On Sales
RPN	Risk Product Number
RSD	Requirements Specification Document
RSS	Requirements Specifications Standard
RTM	Requirements Traceability Matrix
SE	Software Engineering
SEI-CMM	Software Engineering Institute — Capability Maturity Model
SLA	Service Level Agreement
SME	Subject Matter Expert
SOW	Statement Of Work
SPI	Schedule Performance Index
SPMP	Software Project Management Plan
SRS	Software Requirements Specifications

SV	Schedule Variance
SWAG	Simple Wild Ass Guess
SWOT	Strength, Weakness, Opportunity, Threat
T&M	Time and Material
TCPI	To-Complete Performance Index
TPMP	Transition Project Management Plan
UCL	Upper Control Line (Limit)
USL	Upper Specification Limit
V&V	Verification and Validation
WBS	Work Breakdown Structure

CHAPTER ONE

The Need for Project Management in Engineering

1.1. Introduction and Motivation

Engineers design things; that is what we do. Our companies give us a work order, and we are expected to construct an object that satisfies that order, be stewards over the budget allocated to us to do the work, get the work done in a timely manner, and ensure that the final product does no harm but rather delights the end users. We have been doing this for a very long time. Roads and bridges constructed by Roman engineers[a] 2,000 years ago are still in daily use. Europe is dotted with medieval cathedrals that would have been a huge task for any civil engineer today. Yet these structures were constructed over several centuries, with many engineers sequentially involved in the construction. But until the modern age, little concern was given to the financial aspect of a project. The King proclaimed "let it be done", and the engineer started work. There were no plans or best practices in place. The engineer would work through trial and error until either the project was abandoned, completed, or worse, delivered with a serious defect (e.g., the Leaning Tower of Pisa). As the industrial age of the 19th

[a]Note that we are using the word "engineer" in a generic sense. The word "engineer" became copyrighted and mandated by legislation only in the 20th century and only in certain countries such as Canada, UK, USA, etc.

century restructured the modern world, engineers were thrust into the position of being the principal agents of change. With this new-found popularity came the responsibility of protecting the public from any harm that may arise due to their creations. Thus, out of necessity, engineers banded together in their areas of expertise and started to codify "best practices". This was the beginning of the profession of engineering.

Consider bridge-building. In Roman times, the practice of bridge-building was maintained by lessons learned from a trial-and-error process. Using the apprenticeship paradigm, the Romans would pass on their experience from generation to generation. What was good was retained and the bad discarded. In the ancient times, people improved their skills by learning from failure. Failure is the greatest teacher of all engineers. What distinguishes an engineer from the rest of society is that when faced with failure, the non-engineer simply regards the unfortunate event as an Act-of-God (a spiteful God punishing the human for some transgression) and moves on. The engineer sits down and asks, "Why did that fail?" She strives to understand the root cause of the failure and what can be done to avoid it next time. Indeed, the Roman Army motivated their engineers in a most Darwinian way. When the bridge was completed, a legion of soldiers was marched over the bridge in lock-step (they understood resonance even then) with the engineer placed under the bridge! Failure meant death to the miserable engineer. Today, civil engineers meet with a similar fate, only, the end result is not death, but unemployment. If the bridge falls down, the engineer who designed the bridge can never find work again. Parenthetically, bridges are highly visible structures. Not everyone may understand the principles of bridge-building, but even a layman can see when a bridge fails. Contrast this with software construction, where this is seldom true. For the civil engineer, there is no way of covering up the disaster; the failure is there for all to see. The software engineer, on the other hand, can be clandestine in covering up a software breakdown.

We live in the Engineering Age. Look around you as you drive home tonight (in a vehicle that an automotive engineer designed). Everything that you see is touched by the hand of an engineer. If a product fails, the engineer becomes liable for prosecution, even imprisonment. How does the engineer protect himself against the lawyers and, more importantly, how does the public protect itself against unscrupulous engineers?

Consider the edifying example of the evolution of high-pressure steam. When the Englishman James Watt developed the steam engine at the end of the 18th century, he used what he called "low-pressure" steam. He knew that the use of high pressure would cause the boilers of the time to explode, scalding and killing anyone close by with the superheated steam. Moreover the physics of steam was not understood at all, neither was the metallurgical engineering of the steel used in the construction of the boilers. However the promise of much faster devices was a strong incentive to experiment with it. Particularly in the area of steam ships, high-pressure steam meant that the ships would go much faster, especially on trans-oceanic routes. Because of the increase in speed, high-pressure steam became common in the first four decades of the 19th century; so too did hideous accidents as the boilers routinely exploded. An estimated 40,000 people lost their lives, some scalded to death, some blown into crocodile-infested waters. In those days, there was no regulation of engineering as a profession. Anyone could hang up a shingle and declare himself to be an "engineer". Finally the governments of the day intervened and encouraged the best-practicing engineers to establish a list of "best practices" and set a series of examinations that would-be engineers would have to pass before they were issued a license to practice. In Canada, this evolution to a profession began much later (1903) and was not completed until 1933. Today, to call yourself a professional engineer, you must either have graduated from an accredited university or write a series of skill-determining examinations. Thus the public is protected against unscrupulous quacks masquerading as "engineers". In the high-pressure steam case, scientists worked alongside the engineers to improve the quality of boilers (better metallurgy) and to begin to understand the physical properties of steam (leading to the development of thermodynamics and the Carnot cycle). Thus it was a very rare event to have a steam boiler blow up in the 20th century.

When an engineer constructs a product (be it a boiler, bridge, or piece of software), she has to be a steward. The engineer manages the work, its cost, the time to finish the project and above all that, guarantees that the product will be fit for use. How can she do this? In the past, it was a by-guess and by-golly approach. After the Second World War, engineers started to draw up "best practices" for the running of the project to construct the product. The work now becomes a "project", not just an engineering activity. The running of the project involves best practices that are common to ALL projects, not just civil engineering ones. This then is the purpose of the book: to examine the best processes of Project Management and to integrate them into engineering. Project Management in the 21st century is a core competency that any professional engineer must master in order to be successful. It involves the integration of two life cycles: the life cycle of the particular engineering discipline involved and the life cycle of project management. To give the book some specificity, we use the engineering discipline of software engineering as an example, but any branch of engineering could do. We choose software because we are both software engineers and because, of all the areas of engineering needing project management, software engineering is by far the neediest. This is not a random event. Software engineering is the newest branch of engineering. Understanding how Project Management can be integrated into software engineering gives us insight into how it can be adapted to all branches of engineering. Let us see how one of us came to see the light of Project Management in a specific engineering area (software engineering).

1.2. In the Beginning

Software. What exactly is that? That was the question one of us (JMB) asked himself several decades ago. And you are wondering: "what's this got to do with Project Management?" Quite a bit, as we shall see.

I had been educated as an engineering physicist and had a solid understanding of things electrical. I had just been hired by the

Computing Centre of the University of Western Ontario (UWO) and given the mysterious title of "Software Supervisor". The only problem was that I had no idea what "software" was. Being a well-trained engineer, I scurried to my local library and read all I could about "computers". In those days, they were divided into three categories: analog, digital, and hybrid. Then came my second shock. An analog computer was perfectly understandable to an engineer then just as digital simulation programs are to today's engineers. But "digital" (and they did not mean digital, they meant binary)? How could you possibly construct a computer on zeros and ones? And even if you could, what was this thing called "software"? And next Monday I was to start work as UWO's Software Supervisor!

As fate organizes these things, UWO took delivery of two software-related things that week; me and UWO's first mainframe computer, an IBM 7040. You would recognize it today but you would have been stunned by its huge size and pathetic hardware properties. Room-filling, it had about 100K bytes of memory, an integer add time of 16 microseconds, five magnetic tape units, plus a card reader and a line printer. But more importantly, it had Software installed on it. IBSYS, the first commercial Operating System (much like MSDOS) came as part of the system. Attached to it were several application programs, FORTRAN and COBOL compilers, a Loader, an Assembler, and some other packages. Of course, I had no idea what a "compiler" or "operating system" was, let alone "software"!

Thus began my career as a software engineer; a journey that has lasted over 45 years and has tracked closely the evolution of software development. I managed to figure out how IBSYS worked, how compilers worked, and indeed, how to program. My fears of the binary number system were eradicated and I could even give public talks, defining what software was and why it was "soft". I gave courses in FORTRAN, contributed a little in the development of our own ALGOL compiler, and designed the software design courses for our nascent Computer Science program. Life was good.

By now, our fledgling little group had morphed into a semi-professional group composed of hardware and software engineers (although that term was not coined until 1969) led by a professional engineer whom we will call Dilbert. I, of course, was the Software lead and for a couple of years, all went well. I wrote and supervised the construction of application programs that faculty members needed, in addition to maintaining and slightly extending the capabilities of IBSYS. But I did have a nagging concern; my ability to estimate the time necessary to complete a program was very sketchy. When presented with a new project, I would try to compare it against another program that I had written and use that as the basis of estimation. But how does one compare the complexity of two different programs? I would soldier on and when I had the best estimate I could justify, I would double it. Later I was to find out that many other proto-software engineers were doing the same thing; so many that we now call this the Rule-of-Two in estimation (throughout this book, you will find several such "rules". These rules are really rule-of-thumb estimation algorithms that have received extensive empirical validation).

To every practicing engineer, there comes a day in your career when, with blinding clarity, you realize just how inadequate the training is and how little you really know about your area of expertise. Mine came in 1968. Our boss, Dilbert, had a project. We had just taken delivery of Canada's first time-sharing computer, a Digital Equipment PDP-10. But Burroughs, the supplier of the hard drive used for swapping time-shared programs in and out of memory, was a year late in delivering that to us, rendering the PDP-10 mostly useless. Dilbert had a cunning idea. Why not connect the two memories of the 7040 and the PDP-10 together with a home-built bus and have the PDP-10 use the 7040 hard drive? Then the PDP-10, whenever it needed to swap a program out to temporary storage, would interrupt the 7040. The 7040 would checkpoint the current program (save all program status so it could be restarted later, a feature of the 7040's operating system, IBSYS). A special 7040 driver would then accept data from the PDP-10 and store it on the IBM hard drive. When done, the two would disconnect and the

7040 would restart its interrupted program as if nothing had happened. When the PDP-10 needed that program back in its memory, it would reverse the procedure. It was a brilliant plan, brilliant!

Dilbert called a meeting of the three of us (Norm, the hardware director, me, and Dilbert) and we started to plan our attack. Norm went first. Armed with logic diagrams and flow charts, he professionally slogged through 2 hours, describing all of the electrical details of what needed to be done, presenting rough cost estimates and a Gantt Chart illustrating the construction timelines. It was a thing of beauty. Then it was my turn. "Where's your plan, Michael?" Dilbert opined. Tapping my forehead, just like Mozart did in the film *Amadeus*, when Schickenader asked him for the score for the Magic Flute, I replied "it's in my noodle". Mozart really did have a plan; I did not. I should note parenthetically that I had a very low opinion of Dilbert's software knowledge; after all he was only an electrical engineer; I was the software guy (note that I wrote "guy" not "engineer"). I had had several minor software successes to my credit although nothing this complex. Dilbert went red; steam came out of his ears, and he pounded the desk in front of me. "You call yourself an engineer?" he roared. "Here is your PLAN. First you will write out in point form, all of the properties this software will possess, especially in terms of timing constraints. Then you will present these orally to Norm and me. If we don't like what we hear, you'll go back and rework them and cycle until we all accept them. By the way, these properties are called *requirements*. Then you will do a *high-level design* of the software, expressed in flow charts. You will present the high-level design to the two of us and we cycle until we are satisfied that it will meet the requirements. Then you will do a *low-level design* of each module and we will cycle on each one of the modules. When I am happy with a module, you may start *coding* up that module. But each code segment will be *unit-tested* (I had to have that term explained to me). Each will then be cycled as above. Then you will start to *integrate* the basic modules Noah-style (i.e., two by two), complete with documented integration testing, which we will cycle as above. Then you will do a dry-run *system test* and complete that to our satisfaction. Then you will *document* everything that you have done so we can *maintain* the software. Finally, you will describe the entire construction as we have noted above in a complete *project plan*, complete with Gantt charts, which of course we will cycle and finally approve. That, my boy, is engineering."

Had we published Dilbert's plan, the Waterfall Model (see Appendix A) might have been called "Dilbert Model", for that is a basic engineering design format that is suitable for any engineering activity. For me, it was an epiphany. Software can be engineered just like bridges! Indeed, at a NATO conference on software construction the same year, Brian Randall called for the development of "Software Engineering" [1]. However, it would be three decades before this would actually happen. That came in 1998, when the State of Texas officially approved Software Engineering as the newest licensable branch of engineering. The Professional Engineers of Ontario (PEO) followed their lead the next summer and I was able to apply for a P. Eng. For 36 years, I could not apply because I only had one year of "engineering" experience. Then, with the stroke of a pen, I had 36 years experience as a Software Engineer and could now be licensed by the PEO.

What happened in those 36 years? The simple answer is that we have been slowly developing, rejecting, and verifying software processes. Computer Scientists have been researching in the theory of software engineering, and engineers have been applying those principles on a professional basis (that is, engineering the computer science research results). It is instructive to quote David Parnas in giving a public lecture here; "Computer Science is NOT Software Engineering". Perhaps my experience at UWO is also germane. For 34 years, I had been a Professor of Computer Science, teaching "software engineering" principles, among other things. After taking early retirement, I was asked by our Engineering Dean to establish a Software Engineering program in Engineering, not Computer Science. My former colleagues considered me to be a Benedict Arnold, a traitor to my former friends.

I was no such thing. A scientist approaches life examining the basic principles of nature; an engineer takes the useful parts of that research, wraps it in a way that it can be applied in a consistent, safe process that can be taught to others. Interestingly, the same discussion happened a hundred years ago, when Electrical Engineers were birthed out of the womb of Physics. Physicists were outraged; "we invented electricity, it is ours!" they cried. They were quite right. But the engineering of electrical power transmission is something vastly different from theoretical physics. Thank goodness, we have electrical engineers in charge of our electrical transmission lines, not physicists! The same analogy applies to Software Engineering and Computer Science. Each discipline needs the other. Would you really want the software in your automobile's ABS brakes written by a Computer Scientist? Luckily, the argument has cooled off as we now understand our respective areas of expertise.

Now let us look at the Information Technology (IT) revolution from a 10,000 metre view. It all began with hardware. The birthday of the revolution was about the same as mine; the early 1940s, when an American, John Atanasoff, and an Austrian, Konrad Zuze, created machines that we would recognize as computers. The growth in computer hardware has been spectacular. Following Moore's Law, it has been doubling in capacity about every 2 years. Whether one considers the processor, memory, hard drive, or network, the speed of each has been following Moore's Law for decades. For example, a typical microchip in 1980 could add a million instructions per second; in 2005, the same-priced chip could execute half a trillion. In 2009, IBM announced its first petaflop machine (10^{15} floating point operations per second, but using many processors). Just think of how the price of memory sticks has plummeted while their capacity has soared. Similarly, backup drives have, as of this writing, reached a petabyte in size! Network speeds have even surpassed processor/memory speeds. 10 Gigabit Ethernet LANs are common. Wireless connectivity is widespread and those speeds are approaching a gigabit too. All of these disparate technologies are converging together. What we are witnessing is the greatest invention since human speech.

We are in the midst of two revolutions, which is why our world is in constant upheaval. The revolutions are (1) the digital convergence and (2) the fourth turning. The first, we all understand pretty well. All of our binary traffic, whether networks, video, telephony, or networks, all boils down to a stream of interchangeable bits. Everything can be shared. Think of an Apple i-Touch. This unification of information is providing us with a rich set of interconnected applications that we are only starting to understand. The glue that links all of these pieces is Software.[b]

What we are seeing is an Order Of Magnitude (OOM) effect. Consider the evolution from horse travel to automobile travel. A horse averages 5 miles per hour, an auto 50. But it is not just that you go 10 times faster. Everything changes. Vast superhighways must be built, gas stations constructed to supply the new vehicles, and most importantly, we must change the way our society is structured. My grandmother never travelled to Toronto from her home-town of London, Ontario in her life (2 hours by car); I travel an hour a day to work and back, and that is considered a short commute. IT does the OOM change every 6 years! The Webbing of the world is another OOM effect. It is providing us with a sea change of information exchange. The evolution from the internet addressing scheme from IPv4 to IPv6 is instructive. IPv4 has 32 bits of addressing (big enough to give a unique address to almost all of the people in the world). "Why would you want more than 32 bits?" I used to lecture. Because the Web is connecting far more than people. IPv6 has 128 bits, not quite the number of atoms in the solar system but close. Everything can be webbed. Imagine implanting an IP address in each of your socks. No more missing socks! Just ask your laptop to locate the sock and it will respond, "behind the dryer, my friend".

The second revolution is a societal one, described by two historians in their book "The Fourth Turning" [2]. Following the ancient Etruscans, they observe that societies "cycle" every 80 to 100 years. Each cycle is composed of four subcycles. The passing from one subcycle to the next is called a "turning". The four cycles are High,

[b]Christ even foresaw this. In Matthew 5:37 he says "but let your communications be Yea Yea, Nay Nay; for whatsoever is more than these cometh of evil".

Awakening, Unravelling and Crisis. We are now in the Fourth Turning, the passage from unravelling into crisis! Such periods are characterized by chaos, upheaval, and monstrous uncertainty. Sound familiar? Similar examples in recent history were the French Revolution and World War I. Lucky us! We are living through the throes of two violent revolutions.

1.3. Project Management in Software Engineering

Management of software engineering projects is new and immature, compared with management of construction projects, for example. But this is not unexpected; the emergence of any new branch of engineering is fraught with trial and error. Civil engineers have been at the job for three millennia; software engineers only 50 years. But we are catching up fast. It is vital that we do so, because software is becoming ubiquitous. Can you think of any area of human endeavor that is not dependent on software? To paraphrase Sir Wilfred Laurier, the 21st century belongs to Software. It will drive the digital convergence, heralding a world of total connectivity. It is the all-embracing, all-connecting glue that will converge everything. Figure 1.1 illustrates the shifting size of hardware and software.

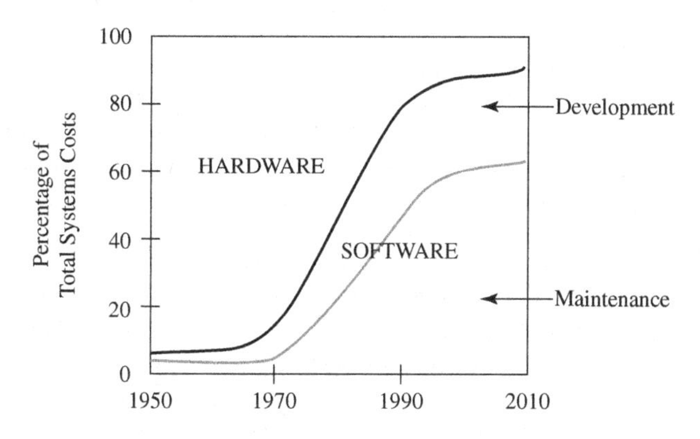

Figure 1.1. Shifting importance of software versus hardware.

In the early days, hardware costs dominated the total price of computer systems. When one of us first came to UWO in the early sixties, software was free; something that just came with the hardware. It was a rude shock to us in the mid-sixties, when IBM "unbundled" software from hardware and charged separately for each. Now we realize that software is very expensive. None of the fantastic OOM advances of hardware apply here. Software does not follow Moore's Law! In fact, software is the dominant cost component of any computer system. Computers themselves (the little chips) are everywhere. Most of them are unseen. They live in electric razors, laser pens and in virtually every device now manufactured. A new car may have a few dozen beavering away on the inside. For every computer you see (as a laptop), there are thousands that are invisible, all being driven by software; all capable of crashing if the software is not right.

But how do we make sure that we get it right? How do we cost it? How do we plan its construction? This is the grand theme of our book but before exploring it, let us consider some of the properties of software.

What IS software anyway? Intuitively it is a set of commands that the computer can follow. Consider the world's smallest program:

```
printf ("Hello World");
```

That little line of C code would be compiled (i.e., expanded) into hundreds of "machine language" instructions that tell the computer processor what to do. Any program in any computer language will follow the same procedure. It then becomes a collection of these commands which in turn are transformed into billions of bits which we call the "program", telling the computer to "do something".

Software's first property is that it is ubiquitous. Your electric razor has 2,000 lines of C code in it. A pacemaker has about 500,000 lines. Think of your cell phone, driven by software. As we will argue, we can engineer software. But unlike all other branches of engineering, software has some very strange properties. Once constructed, we can make infinite copies of it for essentially nothing (contrast that with the making of a car). We are not sure how to engineer it for increased reliability. Normally if engineers are worried about the load-capacity of a bridge, they throw in some more steel. But adding more software to a program normally weakens it and makes it less reliable. If one bit in software is flipped, the airplane is falling out of the sky. In most engineering error analyses, we assume that errors are independent of each other; not in software. They are highly correlated. Software design is based on discrete mathematics, not calculus. Our basic science is computer science, not physics or chemistry.

Software is also very long-lived, mainly because it is so expensive to build. Y2K (the millennium problem of passing from 1999 to 2000, i.e., 99 to 00) illustrated that. Programmers in the 1970s, writing their code when memory was very expensive, chose to use a 2-digit date field for economic reasons. They knew that there would be a problem but they assumed that the software would only be used for a few years and then rewritten. How wrong they were. Software, if it is good, can live for decades and does. Software is also expensive to write. Consider that there are 40,000 nerds on Microsoft's Redmond Ranch, developing and supporting all of those Windows products. That is a big payroll.

Software construction also has other unpleasant properties, besides its costs. As software ages, it is "enhanced" by developers. Additional features are added, which weaken the original integrity of the program. Entropy is alive and well in software construction. Imagine if God decided to "enhance" me by giving me an extra set of arms! Or, perhaps, a set of eyes for the back of my head. I, of course, would lose integrity. Imagine trying to shake hands with another person. Or what could I do if I had two images being flashed through my brain, one forward and one backward? That is what happens to software when it is "enhanced". In fact, it is common to throw an old "enhanced" software product away and rewrite the application from scratch.

Why is it so difficult to manage software construction? To answer this, consider our little story of the "Cops and the Donuts".

We all know that cops like donuts. Suppose I ask you, "How long will it take that group of cops to clean out that donut store?" Of course, you cannot answer the question. Why? First, you do not know how many cops there are. Second, you do not know the eating rate of cops (in donuts per minute which, for the sake of simplicity, we shall assume is constant). Finally, you do not know how many donuts are in the store. Armed with these three figures, you can quickly answer the question. This is the problem with software. We can count the number of software engineers. But how do we measure their productivity rate? What units shall we use? And how do we "size" the problem? If you have these numbers, you can apply project management. If you do not, then programming is a black art and unfit for Project Management. We will have to resort to what we call the "Jolt and Pizza" process. You lock up the software engineers in a room. Give them unlimited supplies of Jolt and Pizza and do not let them out until they have produced the software!

Much research has been done in the recent past to acquire these numbers, as we shall see. Project management has its own set of processes, invariant to the area of construction. What we need to do is to integrate the two life cycles; the common one for Project Management and the specific one to the thing that we are building. We are choosing Software Engineering but you can replace that with any branch of engineering that you like. This ability to estimate and measure both size and productivity rates is key to all branches of engineering. We use software as an example but the ideas flow over to your field of engineering. The amount of software engineering knowledge needed here is minimal.

We begin this book by giving a few success and failure examples of project management both from general engineering and from software. To consider a project to be successful, it has to finish within 20% of the originally estimated time, cost and quality. In examining 3,000 software projects, the CHAOS Group [3] found that only 16% of the existing projects in the industry are successful, while 53% of the projects are being challenged (that is, questioned as to whether they should be allowed to continue or be terminated), and the remaining 31% of the projects actually failed.

An expert in the field, Ed Yourdan, has opined that the average North American software project is one year late and 100% over budget [4]. Even when complete, the American General Accountability Office (GAO) found that only 9% of the software delivered was actually used [5]!

Why is the success rate so low? For starters, a lot of projects begin with inaccurate or non-existent estimates, continue with constantly changing scopes, are constrained by short duration and limited resources and are faced with risks maturing into real problems. In addition, staff assigned to undertake project management roles may not be properly trained in the discipline to undertake the challenge. Most of the time, people merely stumbled into such opportunities. Unfortunately, in the real world, failure will not be tolerated more than once; even though knowledge can only be gained in a hit-or-miss fashion. Trial-and-error effort will result in costly mistakes.

After all, project management is a new profession. For those with kids in your family, go home and ask what they would like to become when they grow up. Some may say a hockey player, a rock star, a doctor, lawyer, teacher, etc. Who would ever say becoming a project manager?

In today's competitive market, there is only one standard for success — excellence. The discipline of project management is no exception. Attempts have been made by various international organizations to standardize the practice of project management. The drive actually comes from rapid changes in high technology that force management and project personnel to re-consider project management techniques. The Project Management Body of Knowledge (PMBOK®) by the Project Management Institute (PMI) is arguably the most influential standard today. PMI has developed certification programs for project management professionals. Although there is no guarantee for project success by following the PMBOK®, there are definitely huge advantages to manage projects in accordance with PMBOK® — the sum of knowledge within the profession of project management. There are other standards such as IEEE, PRINCE, etc. In addition to following some project management standards, the project manager has to gain sufficient application domain knowledge (e.g., how to build a house) and to be aware of the industrial standards and government regulations. He also has to understand the project environment (e.g., cultural, social, political, physical) and possess general management knowledge (e.g., diversity awareness) and skills (e.g., basic accounting, finance management).

1.4. Some Very Expensive Engineering Failures

We have defined a failed project to be one that exceeded either the cost estimates or the time estimates by 20%. In general, we have not done well in software, as we have mentioned. The amount of costs to the North American economy has been estimated to be in the hundreds of billions, yearly.

Let us look now at some specific cost/schedule overruns. Before we look at some software failures, note that there are lots of failures in the non-IT world too. Please note that we use the acronym "IT" (Information Technology) to be more general than just Software Engineering, encompassing not just software but networking, hardware, and so on. Here are some "extreme" examples of engineering project cost overruns.

Spectacular (non-IT) cost and schedule overruns

- Boston's Big Dig — 370% over budget, 320% over schedule

Civil engineering projects are notoriously difficult to predict. If the engineers know the soil conditions well, they can be highly accurate. For Boston's Big Dig (its official name was the Central Artery/Tunnel Project), all the bad things one could imagine happened. It was a megaproject to reroute Interstate 93 from above ground to a 3.5-mile-tunnel underground. It was estimated to cost $2.8 billion (1982 dollars; $6 billion adjusted for inflation) and take 5 years to complete. It took 16 years to complete with a final cost of over $22 billion. Again, the failure was a combination of unknown terrain, poor engineering, and downright fraud.[6]

- Suez Canal — 1,900% over budget

  Ferdinand De Lesseps stunned the world when he built the Suez Canal in 1880. However, the cost overrun was staggering. To be fair to De Lesseps, estimating was in a primitive state at the time. However, it is clear he deliberately underestimated the cost in order to get the project started. This was an example of what we now call a "not-for-profit megaproject". These are common and their proponents invariably overestimate the use of the final product [7].

- Sydney Opera House (SOH) — 1,400% over budget

  Although perhaps the most beautiful public building in the world, the construction of the Sydney Opera House was a project disaster. The product of a brilliant Danish architect, the project was 350% over schedule and 1,400% over budget. The cost overruns were caused by novel construction techniques and by a very arts-unfriendly state premier. Note that even though a project may be a failure in terms of project management, it may well be successful as a human activity. The SOH has become the symbol of Australia, and one can well argue it is the most beautiful building in the world.

- Concorde — 1,100% over budget

  A joint project between France and Great Britain to build the world's first (and only) supersonic aircraft, it was by project management standards, a huge disaster, with cost overruns so extensive, it never achieved profitability. Note though, that it was a superb aircraft.

- Pickering A Rehabilitation (Canada) 300% over budget

  A project to refurbish an aging reactor at Pickering cost Ontario Power Generation three times the estimated cost. It was also 18 months over schedule. This fiasco cost the president of the company his job!

Spectacular IT project failures

Note that it is rare to learn about IT failures. When bridges fall down, all the world sees the failure. When software crashes, the company will hide the fact and we may never learn about the dreadful details. This is unfortunate. Failure is the mother of engineering success. If we know why a project failed, we can identify the root cause and avoid this the next time. This after all is the essence of Engineering. Only when the sorry details are revealed publicly by the Auditor-General or the courts, do we learn the true reasons. Here are some of those cases.

- AAT/CAATs

  Flying in the USA is a dicey proposition. The air traffic control of all major airports is controlled by ancient computers dating from the late 1960s. They are hybrid computers which means that they are half vacuum tubes and half discrete transistors, sporting a memory of 64KB. Yes that is KB. That they have worked so well for so long is a tribute to both the hardware and software people that built and maintained them to this day. In the late 1980s, Congress approved funding for a massive re-write of the system. The company that won the contract was Lorel Corporation. These were the folks, who, when part of IBM's Federal Systems Division, wrote the initial NASA software. That software was so good that we use the term "moon-shot quality" as an adjective to describe software of exceptional quality. Thus expectations were high that they would repeat their success. But after 6 years of muddling and 9 billion dollars being spent, Congress cancelled the project. Lorel had not delivered a single line of working code. Lorel had exemplified what we call the "hero" approach to software development. There being no identifiable process, some genius managed to write a spectacularly successful piece of code. But when she tried to repeat the success, it was a dismal failure.

- CONFIRM

  CONFIRM was a logical attempt to build a bridge too far. The idea was good; to combine the ordering of an air flight with a hotel and car reservation. When you planned your vacation, all three would be ordered at the same time and under the same credit card charge. To build this, a consortium of three companies joined together. They also decided to use a new technology at the time, Object-Oriented Programming, when the technology was largely unknown. Each of the three worked on a different part of the project, supposedly managed by Hilton, the hotel involved. As the project proceeded, each of the three posted spurious completion numbers. Initially, this was not done out of deceit but out of ignorance. If there is no process in place, measuring completion rates is impossible. In fact, for years, software engineers used the Ninety-Percent rule. That is, all software projects at any time were 90% completed. What it meant was that the engineer had no idea whatsoever how long it would take. But he wanted to please the Pointy Haired Boss (PHB) so he said 90%, hoping that the PHB would be happy and leave him alone to complete the work. Eventually, even the most optimistic engineer could not quote the 90% rule so they fudged production figures. Eventually Hilton caught on and cancelled the project, writing off 200 million dollars. Several law suits followed, adding to the monetary carnage.

- Denver International Airport (DIA)

The new Denver Airport was to be the biggest and best ever built in the USA. Construction was started in September 1989 and was to finish 48 months later. It finished 64 months later, with a schedule overrun of 33%. The cost overrun was 70% ($4.8 billion instead of $2.8). The usual suspects caused the problem (strikes, new construction techniques, etc.). But the airport has been a huge success, having the highest on-time arrival rates of any major North American airport. It is perhaps best known in Software Engineering circles because of the automated baggage system. Baggage delivery, as we all know, is a messy proposition with airlines. The idea here was to have real-time computers control the complex delivery from airline to customer (and vice versa). When the system was demonstrated to reporters, they were treated to an edifying sight of bags being ripped open, bags tossed off the conveyor belt, clothes everywhere. The software company concerned had to pay huge penalties until, finally, the whole idea was scrubbed and DIA reverted back to the conventional way of moving baggage. Humans now carry passengers' bags through the Denver International Airport.

- AT&T Network Problems

AT&T is renowned for reliability. For over a hundred years, it has been the epitome of solid, reliable telephonic engineering. Then came 1984. To understand what happened, we need to understand how the long-distance telephone network is constructed. It is much like Internet servers. Huge switches connect local calls to themselves. Then they connect to other huge servers and pass on the long distance calls to the appropriate local switch which completes the call. These huge switches can fail. When they do, they inform all of their connected neighbors that they have gone down and would they please find alternate connections. This worked well until a software upgrade was inserted into all of the switches. The "upgrade" software caused the switch to shut itself down and reboot. But before doing so, it informed all of the connected switches that it was going down. This caused all the immediately connected switches to also inform their neighbors that they were going down and to reboot. The rebooted switches then came up, discovered that the connected switches were down, sent a message to all of them to shut down and promptly rebooted themselves. Thus hundreds of switches went into BoBo-the-Clown mode, going up and down like crazy but completely shutting down the AT&T long distance systems. The outage was almost 16 hours. It is estimated that over $500 million dollars was lost in lost-business opportunities. The main lesson here: always test software upgrades BEFORE deployment.

- Mars Probes

We have all seen the spectacular photos that the Mars Lander has sent back from Mars. It has been a wonderful success for the Jet Propulsion Laboratories (JPL) and their engineers. But it came with a very expensive cost. A few years ago, JPL adopted a new Project Management model; faster, better, cheaper. As we shall shortly see, it is impossible to meet all three criteria and JPL was about to find out the hard way. The first of the two probes was the Mars Orbiter. It was to fly to Mars and orbit the planet, sending back photos to Earth. It positioned itself perfectly in the Martian atmosphere and then all went silent, and stayed silent. A 300 million dollar fiasco. Of course, NASA convened an inquiry panel. That panel found out that there were four major modules in the software. Three treated their measurements in System International units; the fourth, in Imperial. The factor was about 4, enough to silence the Orbiter forever. Here is a blunder that students in Software Engineering (SE) 101 would not make. "Faster, better, cheaper"? Not!

The second probe was the Mars Lander (Version 1). It was to land and send back photos. On board the Lander were two controlling computers, one at the back and one at the front. As the Lander started its descent to Mars, it of course needed control to ensure that the decent was at the proper speed. But the two computers disagreed. One said "speed up, we're going too slow". The other said "slow down, you are going to crash". Unfortunately, the wrong computer had control and the Lander crashed into the Martian surface. Another $300 million gone. This is a standard software engineering problem; the control of multiple computers under disagreement and we know how to solve that problem. But "faster, better, cheaper" dulled the JPL project management senses.

- Ariane V rocket (odometer problem)

It is not just the Americans that have problems. The European Union (EU) is just as likely to fail as anyone else. A common execution-time failure is the "odometer" problem. Think of your car odometer. It likely has 5 digits and can record mileages from 0 to 99,999. What happens when it tries to record the next mile beyond 99,999? It resets to 0! We refer to this as the odometer-rollover problem. The famous Y2K was a spectacular example of that. When it happens in your car, it is amusing. When it happens at 30,000 feet in the computer of a 747, that plane is earth-bound! Why this is common is that any computer has a finite "word size" for representing numbers. When we speak of a 32-bit computer, it means that it can represent integers up to about 4 billion. There comes a point when adding 1 to the number will cause it to reset to zero. Of course, you can program the computer to do something when this happens but we often forget as programmers to cover this error condition.

The Ariane V was such a failure. The model IV was very successful. The software was not upgraded for the version V; it was just copied over. But the hardware on the model V was much much faster than that on the model IV. The word-size on the model V rolled over in about 10 seconds. On its maiden test, the rocket had to be destroyed because it was flying blind. The solution was simple of course; increase the word size.

- Red Cross and Hershey Server Meltdowns

With the advent of the Internet, companies have been building huge, interconnected systems that allow for global sharing of data. But any engineered product has to be performance-tested before being released into the field. Imagine opening a bridge without first doing load-testing on it. But we do this all the time with Web applications. Our favorite web loading example is the American Red Cross server, which kept track of donations for the charity. The day before 9–11, it had collected a few thousand dollars of donations. On that fateful day, the servers crashed miserably after having collected over a million dollars. Who knows how much money was lost because of server unavailability?

Hershey is a chocolate manufacturer. For Hershey, 70% of all annual sales occur in the weeks before Halloween. The IT folks decided to launch a huge Web application. The idea was to tie all of their customers, worldwide, into a centralized server farm and schedule delivery from the factories to the warehouses to the customers. The system would load-level supplies and demands so everyone would get just what they needed and not much more. The idea was magnificent. The execution was horribly flawed. No one load-tested the Web applications, and they launched it at the busiest time of year. The application went belly up, unable to handle the huge load. The losses were staggering; over $700 million in lost sales.

1.5. Some Very Deadly Projects

Any time a project causes death, the project has failed almost by definition. Here are a few of them:

The first collapse of the Quebec bridge: 1907 (75 people killed)

Every Canadian engineering student dreams of the day when she will get her iron ring; that iconic symbol declaring that you are an engineer. The two of us were no exceptions. The "Ritual of the Calling of an Engineer" ceremony that confers the iron ring on soon-to-graduate engineers, was written by a Nobel Laureate (Rudyard Kipling), in response to the greatest Canadian engineering failure of that time, the collapse of the Quebec bridge in 1907. What made the ceremony even more mysterious was that we were told the steel for the rings came from the wreckage of that terrible disaster. Thus, whenever we feel that ring on our finger, we are to remember the enormous trust the public places on ourselves to "do no harm". Failure is the engineer's greatest teacher.

Of course, the ring origin story is an urban legend but a nice one at that. The real story is again most instructive for the nascent engineer. Quebec City is strategically situated at the point where the St. Lawrence River first meets tide and salt. For over a century, people had envisaged a bridge to connect the City to the south shore. But the span necessary to do that was beyond the capacity of bridge builders of that day. Then the Scots spanned the Firth of Forth (about the same span necessary for Quebec) using a new design called the cantilevered bridge. This breathed new life into the idea. This coupled with the upcoming tercentenary of the founding of Quebec, led to the formation of the Quebec Bridge Company (QBC) to do the work (for a much more detailed and fascinating account, see Henry Petroski's *Engineers of Dreams* [8]). QBC, the prime contractor, subcontracted the construction of the superstructure to the Phoenix Bridge Company in

Pennsylvania, who in turn subcontracted the fabrication of the steel components to the Phoenix Iron Company. There was a local engineer on-site but the primary designer was a New Yorker by the name of Theodore Cooper. Mr. Cooper, although in his 70s, was considered one of the best, if not the best, bridge builders in the world. It is ironic that the QBC went state-side to find their engineer, claiming that there was no Canadian qualified enough to do the work.

Work started in 1899. As mentioned, there was considerable pressure to complete by 1907 for the tercentenary celebrations. As the first span of the cantilever slowly reached out over the water, serious deformations of some of the supporting members were noted and a cable was sent off to Cooper in New York. After checking some of his calculations, he realized that he had made a serious mistake. He cabled back to Quebec to halt work immediately. But the cable was either delayed or not opened. On the 29 August 1907, the span collapsed, killing 75 workers and making it the greatest Canadian engineering disaster up to that time. As my colleague Dr Pop-Illiev likes to tell me "doctors kill their patients one at a time; engineers do it by the hundreds". You will note that the consulting engineer never visited the site once. When we look at the photos of the span just before collapse (and we are no civil engineers), the struts look too thin. Had Cooper been there, he would have stopped work immediately.

As my grandmother always used to say "It's an ill wind that blows no good!" The bridge was finished in the end by Canadian engineers. It is still in service today, over 100 years later. It spurred the drive to professionalize engineers, leading to the formations of the Canadian Engineering Accreditation Board (CEAB), and general licensing of engineers. It led to the Ritual and the Iron Ring ceremony and the stress all engineering faculties place on safety and protection for the public.

In software engineering, we have our own culprits. So much is controlled by software that, if a single error manifests itself, the results can be catastrophic. Here are a couple examples of projects that behaved very badly.

The Therac 25 disaster

The Therac-25 was a cancer radiation treatment machine. Formally called a "Cobalt-Bomb", it used radiation to kill cancerous cells. The machine had two modes; one was a direct fire mode where it blasted away at full power. The second invoked a shield which cut the power of the radiation beam to levels "safe" for humans. The 25 was an upgrade from the previous machine, the Therac-21. The latter was highly successful. The older machine was controlled by a Digital Equipment PDP-8, a 12-bit work-horse at the time. Like many machines of that era, its programs were coded in assembler. When the new model came out, the PDP-8 was replaced by Digital's 16-bit computer, the PDP-11. The two machines had no relationship architecturally. The maker of the Therac-25, AECL, hired a junior programmer from a Canadian university and charged him to recode the program in the new assembly language, which he did. He also replaced some of the hardware controls with what he thought were software equivalents.

The new machine was shipped to several sites in North America. What happened next is common with complex engineering failures. The machine worked fine, as long as no typing errors were made on the input console. Under the conditions that the operator was prompted to enter a certain amount of radiation to be administered AND the operator made a typing mistake, the software erroneously shifted the digit by 10 instead of erasing it. Normally when we make typing mistakes, we make several. In this case, the dosage was increased a thousand-fold. There was no way for the technician to realize the terrible error that had been committed. And of course, computers never forget. This happened to 5 people. They were given 10,000 times the dosage, a dosage so high that the shield was removed to permit the excessive amounts of radiation. People in the room reported seeing blue light coming out of the machine and the smell of burning flesh. Those 5 people were radiated to death. Of course, it took a period of time to track down the mistake but by then, several people had been brutally killed.

The error here was in not exhaustively testing all possible error conditions, particularly when replacing hardware check with software ones. Of course, using an unsupervised new-hire to do such critical work was inexcusable.

The patriot missile

During the first war against Iraq, the Americans were very proud of their patriot missile. In contrast, the Iraqis were using an old Soviet technology called the SCUD missile. The Americans scoffed at the SCUDs and used very insulting terms to describe what they considered to be ancient technology. President Bush (the first one) bragged that the Patriots had intercepted over 40 SCUDs. Of course, we all assumed that they were blown out of the sky. Bush even visited the Raytheon plant and presented them with a citation for work well done. Then came the horrible news. A SCUD crashed into a Marine barracks in Saudi Arabia, killing 27 marines. Lucky shot, we all thought. When the war was over, the Israelis, who have a vested interest in things military when they happen in their backyard, informed the Americans that the Patriots were flying blind. By "intercepted", Bush meant that they merely passed each other in the sky. Not one SCUD had been shot down for good reason. It was another example of the odometer roll-over problem. The Patriot had a 24-bit timer driving its navigational system. At the speeds that Patriots fly, it rolled over after about 23 kilometers. After that, it was flying blind! In fact, the missile was initially developed to be a battlefield weapon, not a weapon of long-distance. When the timer was increased to 32-bits, it became the deadly weapon that Bush said it was. Unfortunately, this discovery was too late for those 27 Marines, though [9].

The London ambulance service debacle

Each project fails in its own unique way. In this case, a well-meaning project was sabotaged by stingy governmental officials. The idea was simple enough. There were dozens of hospitals in London's metropolitan area (London England that is). Often an ambulance would deliver a patient to the nearest hospital only to find it full. The driver would then try the next and in many cases had to try several. So the London Ambulance Service decided to let a contract to connect all of the hospitals in an ambulance network that would share information. The ambulance driver only had to connect to the network and find out where the nearest accepting hospital was. Great idea. They then let the contract and took the cheapest response. It was a network of small computers called cutely enough the Apricots. The people writing the software had no large software system experience and the expected happened. With little testing, the whole system went live (to meet contract conditions). Over the weekend, ambulances were misdirected, hospitals which were quite empty, were flagged as full. It was a nightmare and it was estimated that 45 people died because of the ambulance misdirection.

1.6. Some Successful Projects

Confederation bridge (Canada)

The Confederation Bridge project of the Canadian federal government was estimated to cost $1.3 billion and to take two years to build. It came within ½ of 1% in both accounts. An example of a well-designed megaproject, the project was on-time and on-budget. It, of course, had an extensive and obviously competent Project Management Office (PMO). Now we know a lot more about building bridges than constructing software but that is where we hope to be headed.

Vancouver 2010 Olympics and Canada Line (Canada)

These are two separate projects and both were successful. For the Winter 2010 Olympics in Vancouver, British Columbia, Canada, a final audit conducted by PricewaterhouseCoopers released in

December 2010 revealed that operation costs to have been $1.84 billion coming in on budget, resulting in neither surplus nor deficit. Construction of the associated venues also came on budget: $603 million [10].

The Canada Line is a rapid transit line in the Metro Vancouver region, servicing Vancouver, Richmond, and the Vancouver International Airport. It was built as a public-private partnership. When approved in December 2004, the cost was estimated at $1.76 billion. As of March 2009, the entire project was expected to cost $2.05 billion. The project is on budget and ahead of schedule [11]. Opened in August 2009, fifteen weeks ahead of the original schedule, the Canada Line was in service well in advance of the Winter Olympics in February 2010.

LA 1984 Olympics (USA)

Where the ambitious construction for the 1976 games in Montreal and 1980 games in Moscow saddled organizers with expenses greatly in excess of revenue, LA strictly controlled expenses by using existing facilities with two exceptions: a swim stadium and a velodrome paid by corporate sponsors. The 1984 Summer Olympics are often considered the most financially successful modern Olympics [12].

We stop here but we could have covered a book with similar examples.

1.7. The Need for Project Management in Engineering

If one has good project management, the chances of bringing the project to a successful completion increase enormously. It enables the Project Manager to know at all times, what the state of the project is, from the points-of-view of costs, schedule, quality, completion dates, exposure to risks, and maintenance of quality. In short, it is the KEY competency for the 21st century engineer. For the remainder of our book, we describe the best practices of Project Management in an orderly yet complete fashion.

1.8. Exercises

Gather information on the following project management stories. Add your own observations and comments.

- Three Gorges Dam, China
  Make a report on this massive project. Comment whether it was a success or not.

- Shinkansen Bridge, Japan
  Research this and report as to its success or failure.

- The Chunnel
  The Chunnel connects England with France. It was 80% over budget and vastly behind schedule. Analyze it from a project management point of view.

- De-islandizing Denmark
  In the 1990s, one of the great engineering accomplishments was three connections between the two largest Danish islands and a connection to Sweden. It is now possible to drive directly from the German border to Stockholm. But these three projects were way over schedule (rail tunnel from Fynen to Zealand, auto bridge across the same, Oresund bridge from Copenhagen, Denmark, to Malmo, Sweden). Describe the perils of Scandinavian engineering.

- The Panama Canal
  Describe the two attempts to build the Panama Canal. The first by De Lessups was a financial and human disaster. The second, by the Americans, was much more successful.

- Olympics 1976, Montreal, Canada
  Jean Drapeau, Montreal mayor in 1976, famously declared that the Olympics could no more run a deficit than a man could have a baby! Gentlemen beware, he was wrong. Why?

References

1. Naur, P. and Randell, B. Software Engineering: Report of a Conference Sponsored by the NATO Science Committee. Garmisch, Germany: Scientific Affairs Division, NATO, October 1968.
2. Strauss, W. and Howe, N. *The Fourth Turning*. Broadway, 1996.
3. The CHAOS Report 1995. Boston, MA, USA: The Standish Group, 1995.
4. Yourdon, E. *Death March*, 2nd Edn. Prentice Hall, 2003.
5. Recovery Act. GAO-10-807, US Government Accountability Office, 1980.
6. Available at: http://en.wikipedia.org/wiki/Big_Dig.
7. Flyvbjerg, B., Bruzelius, N. and Rothengather, W. *Megaprojects and Risk: An Anatomy of Ambitions*. Cambridge University Press, 2003.
8. Petroski, H. *Engineers of Dreams: Great Bridge Builders and the Spanning of America*. Knopf, 1995.
9. Available at: http://www.historycentral.com/desert_storm/iraqinvades.html.
10. Available at: http://en.wikipedia.org/wiki/2010_Vancouver_Winter_Olympics.
11. Available at: http://en.wikipedia.org/wiki/Canada_Line.
12. Available at: http://en.wikipedia.org/wiki/1984_Summer_Olympics.

CHAPTER TWO

The Engineering Project Management Context

We begin by offering a definition of Project Management. We continue by examining the concept of a life cycle and how it relates to both Project Management and specific engineering disciplines.

Engineering projects do not execute in a vacuum. They are encapsulated in some organizational structure that provides the resources (of all kinds) to run the project. Different functional parts of the organization can contribute specialized knowledge to the project at appropriate times. It is critical for the Project Manager (PM) to understand these relationships and utilize them to her best advantage. Our road map for Chapter 2 is:

- Formal Definition of Project Management
- The Goal of Project Management
- The Standardization of Best Practices of Project Management
- Project Life Cycle and its Phases
- Project Stakeholders
- Organizational Influences
- Professional Ethics

2.1. Formal Definition of Project Management

Let us begin by offering a vague definition of the word "project". According to the Oxford English Dictionary, a project is a noun meaning "an enterprise carefully planned to achieve a particular aim" or "a piece of research by a student". As a verb, it means to "estimate based on present trends", "plan", "extend outwards beyond something else", and "cause to move forward". Essentially, it is a human activity to construct an artifact that will require money and time to complete. Now, note that we have quite a collection of our so-called P-Soup; words related to project that begin with the letter P. We have the following:

- Project
- Process
- Product
- Project Manager
- Project Management
- Program
- Program Manager
- Program Management
- Portfolio
- Portfolio Management

A "Process" is the formal description of the way, using industry best-practices, to construct or execute something. This book describes the process of project management. The "Product" is what the project creates. "Project Management" is the application of knowledge, skills, tools, and techniques to project activities, to meet the product's requirements. The "Project Manager" is the person assigned by the performing organization to achieve project objectives. A "Program" is larger in scope and often comprises several interrelated projects managed in a coordinated way. Related terms are Program Manager and Program Management. A "Portfolio" is a collection of similar but unrelated projects and "Portfolio Management" is the controlling of this set of projects.

There are several different formal definitions of what is meant by "project". The Project Management Institute defines a project to be "a temporary endeavour undertaken to create a unique product, service or result" [1]. The British Standard BS 6079-1 defines a project as "a unique set of co-ordinated activities, with definite starting and finishing points, undertaken by an individual or organization to meet specific objectives within defined schedule, cost and performance parameters" [2]. PRoject IN Controlled Environmnt (PRINCE) defines a project as "a management environment that is created for the purpose of delivering one or more business products according to a specified business case" [3], or "a temporary organization that is needed to produce a unique and pre-defined outcome or result at a pre-specified time using pre-determined resources". Juran defines a project as "a problem scheduled for solution" [4].

Regardless of which definition we use, an activity must have the following five characteristics to be called a true project. We shall use this as our definition of a Project. An activity is a Project if it:

1 is a unique, one-time activity,
2 uses limited resources (people, money, time, etc.),
3 has a precise goal,
4 involves sequenced activities which are partially ordered, and
5 has a timeline with a start and an end point.

Let us look at each of these in sequence.

1. Projects are truly unique. Even if you were to execute the same project twice (as has been suggested in the US military as an appropriate process for military software construction), you would have learned much from the first project that you would use to improve its re-execution. As Heraclitus wrote millennia ago, "you cannot step twice into the same river, for other waters are ever flowing onto you".
2. There is always a pressure to use resources wisely as we never have enough money, never enough time, never enough adequate resources.
3. Projects have a precise goal which can be stated in a single sentence. Beware of umbrella, omnibus projects that contain many goals diluting their original sense of direction and true purpose.
4. Partially ordered means that some activities must be done before others can be started. But some activities have no

relation to others and can be started or not-started at will. For example, suppose that your project is to recarpet and repaint your classroom. The primary activity, A1, which has to be done first, is to "empty the room of its furniture". The second activity, A2, "take out the old carpet" cannot be started until A1 is done. Then the next four activities, A3 "paint the north wall", A4, "paint the south wall", A5 "paint the east wall", and A6 "paint the west wall", can be done in any order. Indeed they could be done sequentially by one resource or done in parallel by four. All of these would have to be completed before we could start A7, "lay the new carpet". Then the last, A8 "restore the furniture" would complete the project.

5. A timeline specifies when the project is to be started and when it is to end. The line will be marked off in time units appropriate to the project (hours, days, weeks, months, years). When the requirements of a project are satisfied, it signifies the completion. (Of course, if a project is prematurely terminated, it ends by either party in default.)

Any putative project can be analyzed against these five characteristics. Many activities are similar to projects but lack one or more of these characteristics. For example, pure research is not a true project as we do not know if the goal can be achieved.

Here are some typical projects:

- designing a new automobile
- building a bridge, house, cathedral
- writing a piece of software
- installing a new feature in a house (pool, bathroom)
- writing a book
- getting a degree

Projects need to be distinguished from continuous processing; that is, repetitive activities that are ongoing and never stop. Such examples include:

- claims processing
- assembling cars at a plant

- cooking in a restaurant
- maintenance
- Fedex delivery

Note that some aspects of these ongoing activities can be viewed as projects. Consider Software Maintenance. It lasts as long as one uses the software; it is indeed ongoing. But a single activity (correct this bug, install that upgrade) can be considered as a project. Activities such as maintenance are often called "programs".

2.2. The Goal of Project Management

We can state this simply. The goal of Project Management is threefold — to complete the project:

1. on time
2. on cost
3. on scope

This is often likened to a three-legged stool or expressed as a triangle (often called the Iron Triangle, as shown in Figure 2.1). But note that this is an unattainable goal. As the famous oil-well extinguisher expert, Red Adair, expressed it to the Emir of Kuwait upon presenting him with his bill for putting out 600 oil wells fired by the Iraqis during the first Gulf War, "you can have any 2 but not all

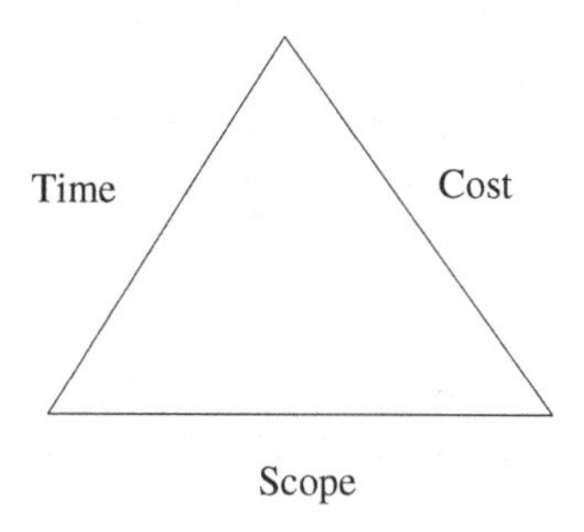

Figure 2.1. Triple constraint.

three. I did it fast and I did it well so here is my huge bill." Note that quality/performance is subsumed in Scope.

How do we achieve this goal? We will succeed by examining the ten fundamental phases of project management (as specified by PMI) and look at best practices which will satisfy the requirements of each of the ten phases. The ten phases include integration, scope, time, cost, quality, human resource, communications, risk, procurement and stakeholder. Indeed, these will occupy us over the next few hundred pages of our book. Note that the PRINCE process model consists of eight distinctive processes that are similar, but not identical, to these ten phases. The eight processes are organization, planning, control, stages, management of risk, quality in a project environment, configuration management, and change control.

Project Management, as a structured activity, has its own life cycle; ten distinct phases that we must execute. This life cycle and these phases are common to all of project management. But the service that we are building or the product that we are constructing; has its own life cycle too. The secret of successful Project Management is to integrate the two distinct life cycles. We will use the standard Waterfall Model (see Appendix A) of software construction, which is the basic engineering model of construction, as our model but you can insert your own flavor of engineering or indeed, of baking a cake. The approach will be the same.

One last comment on Project Management: it is successful if the product "meets" the requirements. In today's world, the motto for success, as expressed by many executives, is to "exceed" customer expectations. Unfortunately, from a pure project management perspective, exceeding customer expectations implies additional requirements being imposed on the project. If these are part of the initial scope definitions, the project can be planned and run accordingly. Otherwise, they are indeed imposing additional effort on the project, which may turn into risks in meeting the original requirements.

2.3. The Standardization of Best Practices of Project Management

As a professional discipline, Project Management was an outgrowth of the frenetic activity of the Second World War, especially by the American forces. That war effort required enormous amounts of organization to deliver material to the fronts. One of our fathers fought alongside the Americans in Sicily, and he reported being astonished by the American logistics. Those soldiers had ice-cream delivered to them in 45°C heat in the middle of a raging war! Imagine the effort necessary to do that.

When the war was over, some forward-looking engineers decided to codify the process of project management. They founded the Project Management Institute (PMI) to oversee the creation of the new profession. The PMI, based in Virginia, has codified the basic components of Project Management, those necessary (but not sufficient) to execute a successful project. These components are detailed in *Project Management Body of Knowledge*, affectionately called the PMBOK® (pronounced PIM-bock). It is currently in its 5th edition but the 6th may be available when you read this. For the rest of the book, we shall follow the PMBOK® closely but not slavishly.

The PMBOK® is organized on the IPO model. IPO stands for INPUT PROCESS OUTPUT and is the standard structuring paradigm of many engineering processes (see Figure 2.2). The Waterfall Model of software development is an IPO. The UNIX operating system is another engineered artifact that is structured on the IPO model.

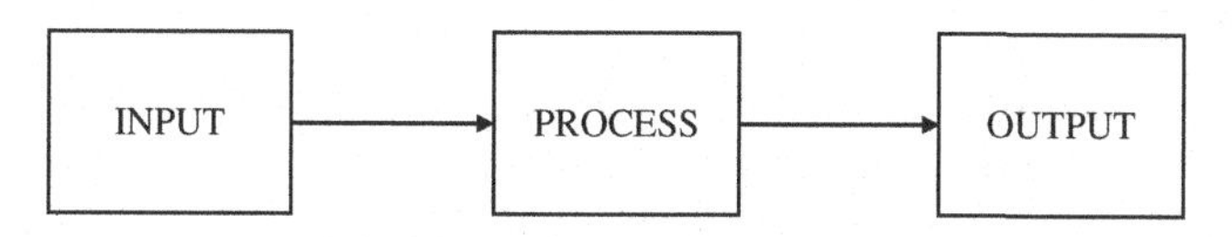

Figure 2.2. The IPO model.

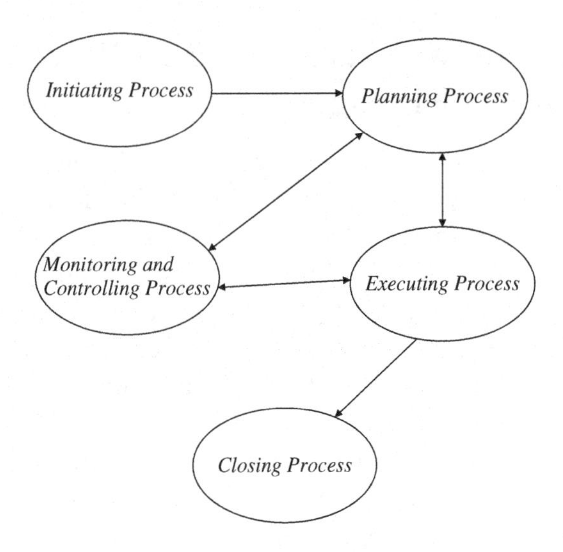

Figure 2.3. Process group linkage.

The PMBOK® defines five process groups, as shown in Figure 2.3.

Initiating processes

The Initiating Processes are concerned with establishing the project as a legal entity, assigning a PM, defining the Charter, identifying stakeholders, and beginning the process of running the project. A new project is typically initiated through the cycle of needs emergence, recognition, and articulation, followed by the definition of formal functional and technical specifications.

Planning processes

Planning is the most important process of the five; leading to the definition of the project management plan (PMP, see Chapter 3). If the plan is right, the next processes will be right too. This is the process by which we select the best of several alternatives for implementing the work.

Executing processes

The executing processes involve co-ordination with people and other resources to execute the PMP in order to accomplish project requirements. Note that the bulk of the budget for a project will be expended in this phase. Project success or failure is very much tied to people (see Chapter 9).

Monitoring and controlling processes

The monitoring and controlling processes of the project define and measure how well we do to create the product. These are the processes where we validate work by checking with the plan. Project performance is observed and measured to identify variance from the PMP. Potential problems are identified and corrective actions can be taken to control execution of the project. Feedback is also provided between project phases. Note the feedback loops into planning if authorized changes must be made.

Closing processes

Finally the closing processes wrap up the work. These processes formally terminate all activities of a project or project phase. The final product will be handed off for verification and signed off by stakeholders for formal acceptance of the completed project scope and associated deliverables. Note that success of the project is dependent on customer satisfaction. Any cancelled project will also be closed. Typical administrative closure activities, regardless of project cancellation, include project diagnosis, evaluation, lessons learned, process assets update, and procurement closeout.

This is an example of a PDCA quality cycle which we will cover in more detail in Chapter 8. PDCA stands for Plan-Do-Check-Act. The PMBOK® illustrates this as shown in Figure 2.4 [5].

Now that we have introduced informally the concepts of what a project is and what project management is, the rest of the book will expand and explain the details of the ten processes that constitute project management.

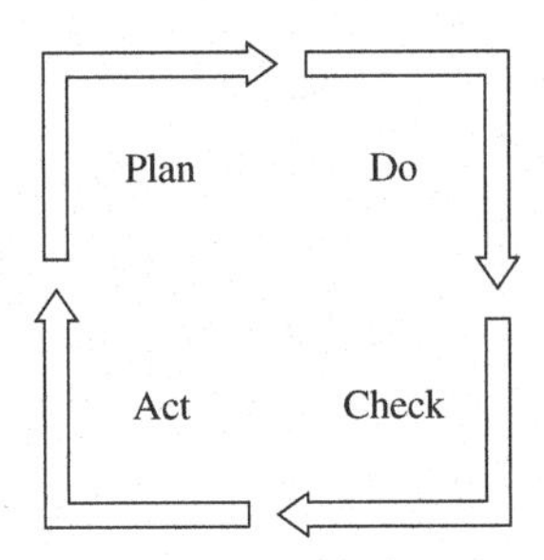

Figure 2.4. The PDCA cycle.

2.4. Project Life Cycle and its Phases

A Project Phase is a "collection of logically related project activities" that culminates in a deliverable. A "deliverable" is a tangible, verifiable work product. Examples are work artifacts such as a feasibility study, requirements definition, a piece of code, etc. We can further differentiate between project management deliverables (such as the project plan) and product deliverables, such as the high-level design document. The important piece is that there is something physical that can be touched, read, understood, and qualified. Phases are normally executed sequentially, from the beginning of the project to the end, but some overlap is permissible, provided that the risks are deemed acceptable. Project management phases are the fundamental components of the project management life cycle. It is independent of the product or service that we are building. As the ISO 9000 Quality Standards are independent of the particular industry applying them, so too is the project management life cycle adequate for all projects. The product life cycle is completely dependent on the product that we build. Again, using the example of software construction, we might use the Waterfall model, the Evolutionary model, the Agile model, the V model, or a multitude of other life cycles available. It is typical of new engineering disciplines to have different life cycles or approaches to solve basic problems. Note the heated arguments of whether one should use alternating or direct current for long-distance electrical transmission. That argument has been running for 100 years and still has not been settled definitely. The Project Life Cycle is a combination of the Project Management Life Cycle and the Product Life Cycle. These phases will be defined by the controlling organization and if that organization is mature, the Life Cycle phases will be defined by a methodology (a document describing "the way we run projects around here").

Phases will have sub-phases and these also will have deliverables. To make this concrete, let us use a simple example of the construction of a critical software application from Nuclear Engineering by the name of "SPENTFUEL".

2.4.1. SPENTFUEL (spent nuclear radiation fuel)

The storage of spent nuclear radiation fuel bundles is a pressing ecological problem for which there is no current solution that would satisfy all societal stakeholders. That solution may come one day. For example, it has been suggested that we load the nasty material onto a rocket and send the rocket off to the Sun. The Sun would never notice a smidgen of radiation leftovers and we would have solved a potentially devastating earthly problem. If we were to find such a solution or a similar one, there would be the difficult problem of locating all of the stored material in North America (NA), as the burial of spent waste is controlled by many separate, unconnected agencies. If there were a central repository containing the location and amount stored for all nuclear burial sites, finding the spent fuel would be a relatively simple operation. The goal of SPENTFUEL then is to build this database.

SPENTFUEL, formally, is a nuclear engineering software project to develop a database to record the location, agency in charge, burial details, and amounts of all of the spent radiation burials in NA. This database would be accessible to all nuclear plants in NA and of course, to appropriate regulatory officials. The idea is that if we ever find a permanent solution to the problem, all of the spent radiation could be collected and disposed. SPENTFUEL will tell us where they are located. You have just been appointed PM to write the Project Management Plan (PMP) for SPENTFUEL. Throughout the text, we will use SPENTFUEL to illustrate project management subphase construction details, paving the way to constructing a complete Project Plan by the end of the book.

Assume that you work for Vacux, a typical old-fashioned company that does general software contract work for the nuclear industry. You work in its IT department, which has announced that it will follow the PMBOK precisely in its project management plans. The rest of Vacux thinks PMI is a Greek prefix. You are the PM and report to the manager, who controls the money. Vacux also has a Quality Control department. Within IT, you have a Database Functional Silo from which you may ask for technical support. Now the funding for our project is to come from the Department of National Defense (DND). The activities will also be vetted by the Canadian Nuclear Association (CNA) to ensure compliance with relevant nuclear standards and regulations.

The software has to be of ultra-high quality. We will assume that we must do at least the following process steps (see Appendix A):

- Requirements
- Design
- Coding
- Testing
- Integration
- Deployment

We will also assume that there are two major independent modules: the control module and the database module. Each module is approximately the same size. Each phase will be reviewed to ensure appropriate quality.

2.4.2. Project gates

The conclusion of each phase, where the deliverable(s) are presented and accepted (that is, the PM achieves "sign-off"), is called various names, namely the *phase exit, gate, kill point*, or *end stage assessment*. We shall use the term "gate". Its purpose is to determine whether money should be allocated to proceed to the next phase. There will be review(s) of the deliverable(s), review(s) of quality, discussion of found defects, proof of defect removal, and so on. Each gate is defined in the Project Life Cycle.

What happens if you do not have gates? This has been the problem in software construction. For decades, good organizations would run their projects using the waterfall model. The PM would give his best estimate of the cost and length of the project. The Organization would transfer the total amount of the project into the project budget, and work would begin. But note that there is no way of telling if the work had been done appropriately. The final day of the project comes along and, lo and behold, it is far from completed.

The use of a Gate places the onus on the PM to prove that it is still profitable to keep going with the project. This may seem like a subtle difference and it is, but it is critical. In most hierarchical organizations, power resides in upper levels; the higher up you go, the more power you have. When a project is established by its charter (more later), there will be a project sponsor who will champion the project to the death. We call these projects "vampire" projects. They are forever doomed to failure but will keep sucking the financial blood out of the organization with no hope of salvation. Not even garlic will help. Now the way to avoid this is to have gating. This places the onus on the PM to earn the money to progress to the next phase. If you have executed the current phase well, funds will flow to the project to continue to the next gate. If you have not done well, no money will flow, and the project will automatically be cancelled. No one is personally responsible, except possibly the PM.

2.4.3. Characteristics of phases and life cycles

Each phase (and subphase) is demarcated by a deliverable. Recall that a *deliverable* is a tangible, verifiable work product, such as a feasibility study, requirements definition document, piece of code, etc. Deliverables are key components in our control of the execution of projects. They are project-created artifacts that a qualified person can decide whether they are acceptable or are not acceptable. This determination is made at the specified gates. In addition to accepting or rejecting the deliverables, other gate activities include:

- prove that it is still viable to fund the next phase of the project,
- review quality,
- detect, report and correct defects, and
- record and publish project cost and time metrics.

We use the Life Cycle model in many aspects of our lives. Think of a newborn child. We have in mind a life cycle development for all children. By the first year, they should be talking, by 18 months, walking, by 12 years, reaching puberty, and by 81 years, dying. As the child grows, we measure each milestone (or gate!) compared with the normal evolution of children. If the child is advanced in development, we might conclude he or she is a genius, a musician, an athlete, an artist, etc. If the child lags in development, we might conclude there is something wrong and do further tests to see if we can correct the perceived trouble. Life cycles are used to compare our particular activity (or child) against established norms of existing best practices.

What kind of life cycles do we have in project management? The two most important are the Project Life Cycle and the Product Life Cycle. Once we define these two, we can use them to:

- determine the complete set of activities to be accomplished,
- show that we are progressing according to plan,
- estimate future costs-at-completion based on completed work,
- estimate project length based on completed work, and
- illustrate that we are maintaining proper quality levels.

There is another extremely important reason to use a life cycle. In these litigious times, project failure will often lead to lawsuits. The PM then becomes accused of professional misconduct, which if

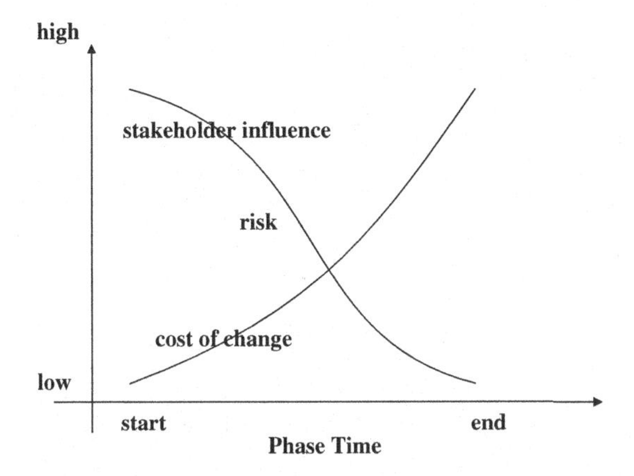

Figure 2.5. Stakeholder influence, risk and cost of change over phase time.

proved, could result in a huge monetary charge against the organization (and you can imagine what will happen to the PM). If the PM can show that he was following best practices, he will be insulated from prosecution and financial charges against the project.

Life cycle phases have certain common characteristics. At the start of the phase, the risk of failure is the highest and declines exponentially to zero as we reach the end of the phase. Figure 2.5 shows this graphically [6].

Similarly, the stakeholders' ability to influence the project and enforce change requests during the phase follows the same curve as Risk in the above diagram.

On the other hand, the cost of change is low at the start of the phase and increases drastically towards the end.

2.4.4. Typical life cycles and their measurement

When we look at the application of effort in completing the phase work, we note that it follows a Rayleigh Curve (see Figure 2.6). The cost or staffing hours, is small at the beginning, rises steeply to a maximum and then declines to zero.

It was the pioneering business researcher, Frederick Taylor, who first noticed that effort (or even defects) over time in business activities tended to follow a Rayleigh Curve. Rayleigh

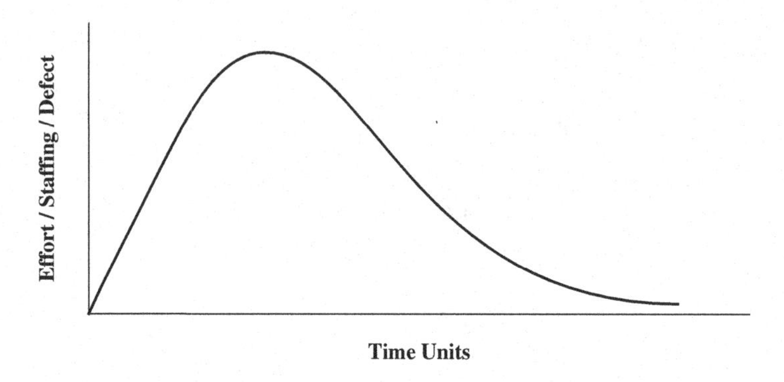

Figure 2.6. A typical Rayleigh Curve.

curves were first described by the famous 19th century physicists, Lord Rayleigh and Max Plank, in an attempt to solve the black body radiation problem. The amount of reflected radiation off a black body, plotted against wavelength, produced the above curve. Lord Rayleigh had a theory that matched the left end of the curve but it blew up at the right end. He opined that if one could come up with a unified theory, physics would then be a closed subject, with everything known. The solution to the problem came in 1900, when Plank introduced his quantum theory which led to quantum mechanics and introduced the era of nuclear physics and relativity (and the rest as they say is history!).

But the Rayleigh Curve does describe the application of effort to complete a phase of work, any work. So much so that we can use it to build estimation models for project execution. A Rayleigh curve is a special example of a Weibull distribution. It has the odd property that 40% of the area lies to the left of the peak and the peak comes at one-third the length of the project. Thus in estimating the work necessary to do the phase, a Rayleigh Curve is a good place to start. The cumulative total, also called the S-curve, is shown in Figure 2.7.

Then estimating the effort of the entire project is a question of summing each phase effort curve to get a project effort curve. Figure 2.8 is such an example.

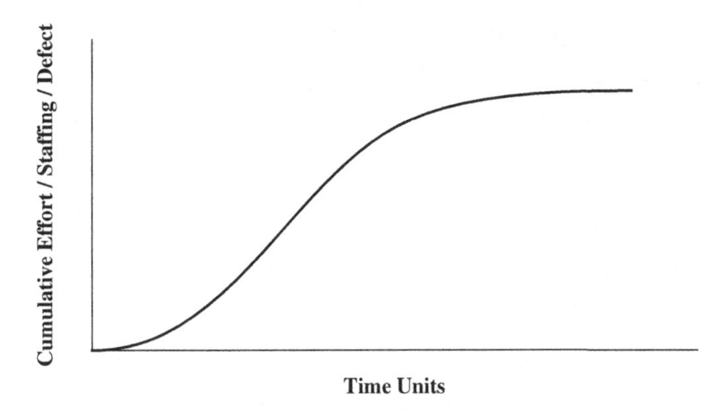

Figure 2.7. The S-curve.

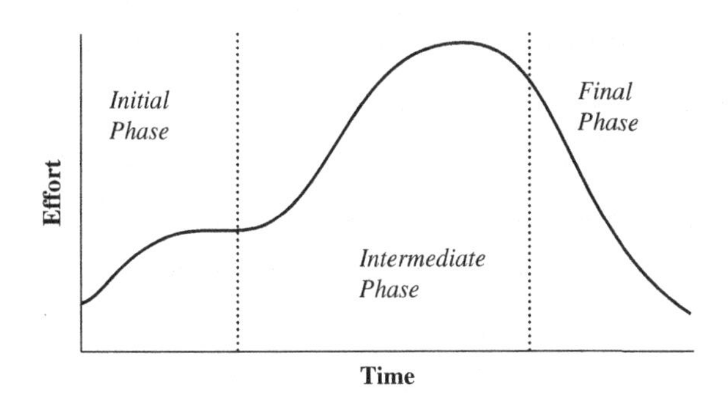

Figure 2.8. Project effort curve.

2.5. Project Stakeholders

The Oxford English Dictionary gives this general definition for *stakeholder*: "a person who has an interest or concern in something, especially a business". The term is also used in mining. When a prospector finds an ore discovery, she stakes (literally with a wooden peg) a prescribed area and files a claim with the Government. When that claim is accepted, she has a fixed number of years to develop the property as a mine and it is her exclusive right; no one else can do so. We call this person the *stakeholder* of that piece of land. In Project Management, a *stakeholder* is defined to be a "person or organization who is actively involved in the project or whose interests may be affected, positively or negatively, by the project or its completion". That is a very broad definition. But it is critical for the PM to identify ALL of the stakeholders and get them involved in the project. Here, we quote the PRINCE project organization, which is based on the customer/supplier environment with three project interests: business, user, and supplier. Some examples of stakeholders include:

- project manager
- team members
- customers
- clients
- performing organization
- sponsors
- project management team/office

We can categorize them in terms of how they affect or are affected by the project. Such categorizations include:

- internal/external
- owners/finders
- sellers and contractors
- families of team members
- government agencies
- media outlets
- individual citizens
- lobbying groups

Similar to the four layers of project organization structure defined by PRINCE that undertake the direction of the project, day-to-day management, team management, together with the team members themselves, we can illustrate the relationships between various stakeholders in Figure 2.9.

We record stakeholders information and details in the Stakeholder Register (see Chapter 13).

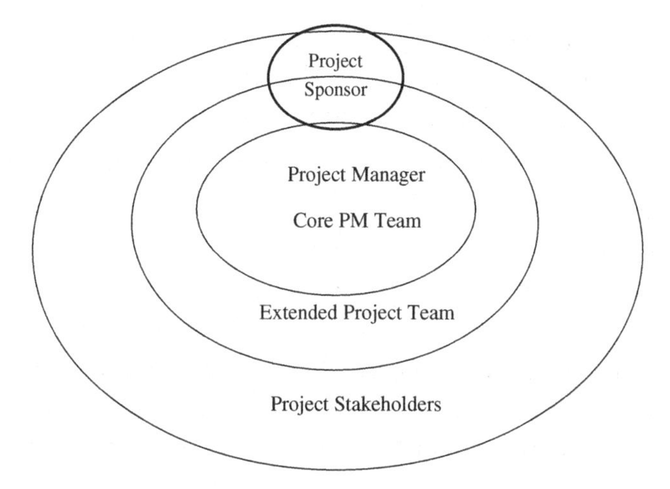

Figure 2.9. The project and its stakeholders.

2.5.1. SPENTFUEL stakeholders

Let us look at possible stakeholders for SPENTFUEL as a typical example. Of course, we will have the Project Team and the PM. We will also have the following:

- manager for Vacux as our organizational contact
- average citizens
- "ADM" = Assistant Deputy Minister (for the two government organizations)
- Vacux BoD = Board of Directors of Vacux
- Parliament — ultimate supplier of funds and justification of project

Now we need to identify all of our stakeholders and ask ourselves, what would they want to know about the project? Following Smith [7], we can use the following process to identify project stakeholders.

- Identify project stakeholders.
- Identify stakeholders interests, impact and relative priority.

- Access stakeholders for importance and influence.
- Outline assumptions and risks.
- Define stakeholders participation.

Now we ask the questions "what would make each stakeholder happy? What are their primary expectations?"

- List the expectations as completely as you can and prioritize them.
- Weight according to Importance-Influence diagram (Table 2.1).

We might end up with a stakeholders diagram such as Figure 2.10.

Now we need to populate each of the ovals with a person's name. It must be a human and you must be able to communicate with him or her. Next, fill in the corresponding impact/influence table.

2.6. Organizational Influences

We need to understand the nature of the organization in which we are embedded. How risk-adverse is the organization? How flexible is it? Is it rigidly hierarchical? Before we can form our project team, we must know how the organization treats and respects projects.

Table 2.1. The stakeholder impact-influence table.

Stakeholders	Interests	Project Impact	Project Importance	Priority

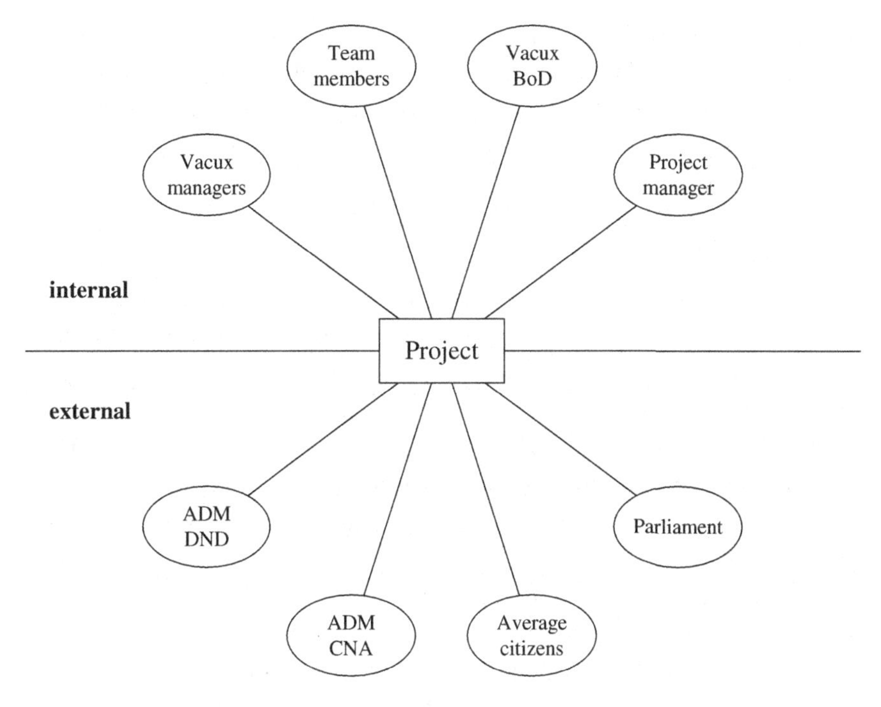

Figure 2.10. SPENTFUEL's stakeholders.

Understanding how organizations function (or don't) has been a problem of close observers for centuries. Look at this example.

> *We trained hard... but it seemed that every time we were beginning to form into teams, we would be reorganized... I was to learn later in life that we tend to meet any new situation by reorganizing; and a wonderful method it can be for creating the illusion of progress while producing confusion, inefficiency, and demoralization.*
>
> Petronius Arbiter (60 Ad)

What a good quote. Unfortunately, it is also an urban legend. The oldest reference to it is about 1975 and was likely authored by a recent NASA engineer, not an old Roman. But like all urban legends, it contains a good piece of truth.

The project/program manager is an individual providing leadership to and acting as the single point of contact for the project/program. She is responsible for the overall project/program and its artifacts (deliverables) via planning, organizing, leading, and controlling. Her roles include those of an integrator, communicator, team leader, decision maker, and climate creator. A common question raised by people is how technical should a project/program manager be? Well, not too technical, although sufficient domain knowledge needs to be possessed by her and the processes or procedures to be followed must be understood in order to produce the artifacts (deliverables) for the project/program. She is like the conductor of an orchestra who does not need to be an expert of any musical instrument, but needs to know how an instrument is played in accordance with the prescribed notes or keys, as well as the order that various notes or keys are put together by various instruments for the whole composition.

A common problem faced by newly appointed PM is the ongoing engagement of technical responsibilities. It is not unusual for him to still undertake one or two designs or programming assignments (part time), while learning and undertaking project management responsibilities (part time). This would be a huge mistake as it poses high risk to the project and the career of the individual! Someone learning and undertaking project management responsibilities for the first time should focus on this very important new assignment, as it should take priority over anything else. The rest of the team, say 10 team members, will be waiting for directions from him and availability of certain project management artifacts (see Chapter 3) for project execution. From time to time, the first-time PM will be struggling with missing the dates of own assigned technical tasks, as well as the management tasks and artifacts that are going to affect many others in the organization!

According to the PMBOK, there are six types of organizations, namely:

1. functional
2. weak matrix

3. balanced matrix
4. strong matrix
5. projectized
6. composite

2.6.1. Functional organization

Functional organizations are characterized by being structured into functional silos, as shown with the SPENTFUEL example in Figure 2.11.

This structure is common, as it provides for easy and efficient management of technical specialists. Resources are centralized and can be efficiently allocated to optimize the function. Knowledge is also centralized, and there will be continuous building of specific expertise within each function. There is a clear career path for all employees. For example, in the database group, staff will be database administrators with various seniority and titles.

There are significant disadvantages from the project's point of view. There is no PM, and cross-functional project engagement is difficult. Any staff who leads a project has no authority. The functional manager is effectively the controller of the project with total authority. She is acting as the technical expert,

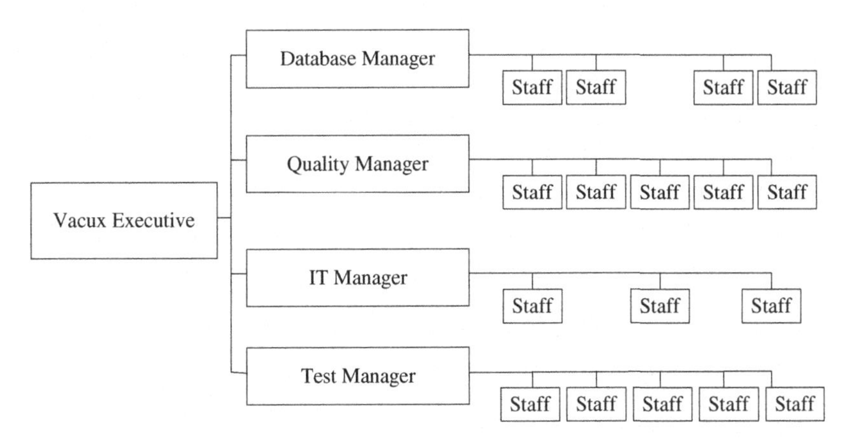

Figure 2.11. Functional organization structure for SPENTFUEL.

resource manager, and project manager, and may become the bottleneck for decision making. The needs of the silo will always come before that of the project. Despite insufficient work, underutilization of staff may not be noticed. There may be little or no opportunity for transferring of staff for career development should the function manager decide to keep the top performers.

2.6.2. Matrix organization

The next four organizations are variants of a matrix approach. The matrix approach tries to combine the functional with the projectized view. These are the forms of matrix organizations (Figure 2.12):

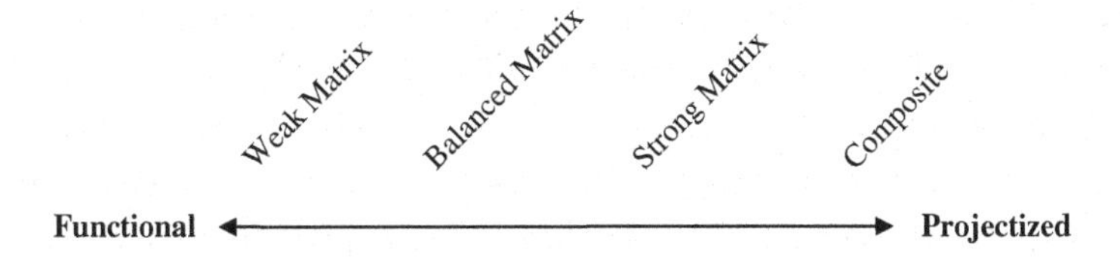

Figure 2.12. Types of organizations and their relationship to project management.

In a matrix organization, cross-functional staff engagement is involved in projects. The advantages of matrix structures are:

- they provide highly visible objectives,
- project management control is greatly improved,
- more support can be provided from the functional silos,
- maximum use of scarce resources will result,
- better coordination of resources will occur, and
- project members have a "home".

Disadvantages include the following:

- the work may not be cost-effective (too many leaders involved),
- team members have two bosses,
- the interactions are more complex to monitor and control,

- resource allocation is more difficult, and
- there is higher potential for conflict and/or duplication of effort.

For a weak matrix organization (see solid ellipse in Figure 2.13), staff will be provided by the function managers of the cooperating functions to work temporarily on a short-term project. No staff has total authority for the project while one of the function managers will be accountable for the project. One of the staff may become a project facilitator (coordinator or expediter) and provide leadership across staff of the other functions.

For a balanced matrix organization (see broken ellipse in Figure 2.13), one of the staff will indeed be a PM, with project management responsibility. However, the PM still reports to the function manager who is accountable for the project. Issues seen by the other functions or different view points than the function and project managers may not be promptly observed or reported.

For a strong matrix organization (see dotted ellipse in Figure 2.13), the PMs report to the Manager of the Project

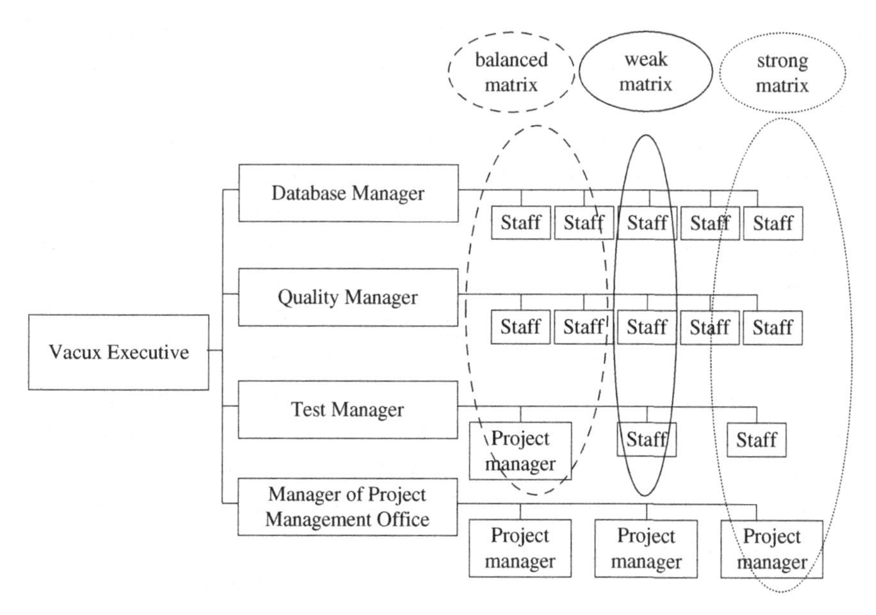

Figure 2.13. Matrix organization structure for SPENTFUEL.

Management Office (PMO), who reports to a top executive. The PMO is a central organization that oversees the practice of project management. PMO sets the standards for project management, evaluates the performance of PMs, prescribes the career path and training for current and potential PMs. In general, this is the ideal mechanism for running projects and for human resource management in organizations. The advantages include a unified and standardized approach to project management that can be tailored for the whole organization with an unbiased view of each project's status and projected completion date. The disadvantages include potentially frequent change of projects for the PMs, which may limit their domain expertise. Conflicts may arise from human resource issues as the PMs have no direct influence over staff in their projects (without any formal reporting relationship).

2.6.3. Projectized organization

The projectized organization (as shown in Figure 2.14) is a modern structure to manage human resources. The PMs report to an executive and they have total authorities for their projects. All activities are taken in a project setting. Once a project is sanctioned, the PM will acquire the necessary staff with domain expertise from the resource pool. During the project, he is responsible for the evaluation of resources in the project. When the project ends, the resources are released to the pool, and he will look for opportunities for new projects.

Naturally the atmosphere will be more efficient from the project management point-of-view. There will be increased ability in project management. There will be increased loyalty to the project and the chances of success will be much greater. There will be full utilization of resources and increased job satisfaction. Additional advantages of this organization are ease of getting on new projects for those who are keen to pick up different skill-sets, and ease of performance evaluation as the top performers are always evident in multiple projects.

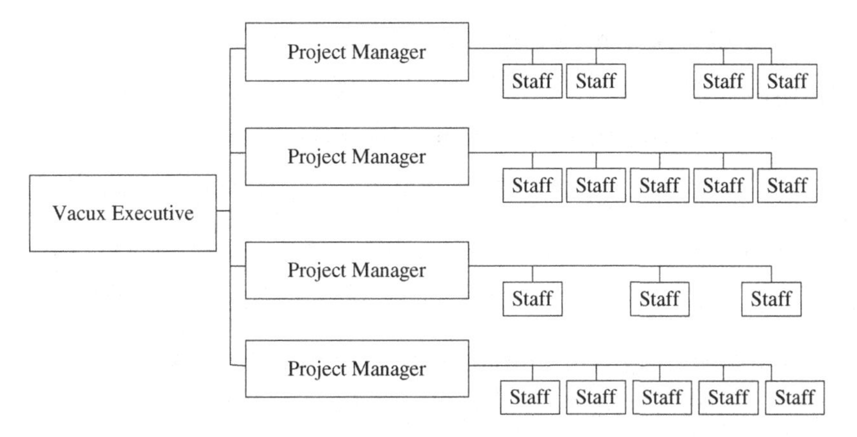

Figure 2.14. Projectized organization structure for SPENTFUEL.

The disadvantages are constant change of projects for PMs and staff which may limit their technical ability in specific engineering disciplines. For those individuals who are highly adaptable and are identified as stars, they may be in high demand, and every project manager would want them on her team. Project priorities may become the source of conflict. There is no "home" to return to when the project is over and loyalty to the organization will be reduced. "Career path" to the top is less clear. Due to the nature of the job, the staff may in fact choose to leave the organization and work as high-paid independent contractors instead.

2.6.4. Final words on organization structure

The PMBOK suggested the Composite Organization, where a project team may be created in one of the matrix forms, to handle a specific "hot spot" sub-project where tight control is necessary.

What type of organization is your firm? There is little chance you can select the form you might like but understanding the general morphology gives you greater insight into the good and bad aspects of that particular choice. We talk more about human resource management in Chapter 9.

2.7. Professional Ethics

Executives love to put up slogans in the work place to reinforce implementation of their visions for the betterment of the organization. These slogans are usually summarized as easy-to-remember acronyms, and they change yearly or every few years or whenever new executives take office. We have encountered numerous ones throughout our career, but one of the slogans that is worth noting carries five "E"s. The first four "E"s are Envision, Energize, Execute, and Edge. An employee is expected to Envision how to map a job to reinforce the organization's visions, and be always Energized to Execute daily assignments and activities with breeding/leading Edge. These four "E"s are governed by the biggest "E" and that is ETHICS. We are sure you agree with this. We cover Ethics and Engineering Professionalism in detail in Chapter 14. You will see how unethical behaviour can cause organizations enormous costs and wasted time on legal issues, not to mention the besmirching of the firm's good name.

As a PMI PMP,[a] one has to support and adhere to the ethical responsibilities described in the PMI PMP Code of Professional Conduct [8]. The code of conduct is divided into two parts: responsibilities to the profession and responsibilities to customers and the public. Being a good citizen alone is not good enough — a PMI certificant is responsible for advancement of the profession too.

Within the profession of engineering, an engineer is typically licensed by an association of professional/certified engineers. To qualify as a professional/certified engineer, one has to have achieved certain engineering education pre-requisites and some years of professional experience, and to pass certification examinations — one of which is Ethics [9]. While practicing as a professional/certified engineer, compliance is expected with applicable statutes, regulations,

[a]PMI has a worldwide certification program (see Appendix G) for project management professionals from all industries. Upon satisfying the education and experience requirements, and passing the certification examination, the designation of Project Management Professional (PMP), or other variations, will be granted.

standards, codes, bylaws, and rules. The code of ethics is a guide to the professional as to the personal conduct in fairness and loyalty to the customers, employers and other professionals. It also governs the engineer's fidelity to public needs and devotion to high ideals of personal honour and professional integrity.

2.8. Multiple Choice Questions

1. Cost of change is the highest on the project during:
   (a) Planning
   (b) Execution
   (c) Control
   (d) Termination

2. Which of the following is not a valid description a project?
   (a) A temporary endeavor undertaken to create a product
   (b) Has a defined start and end point, and specific objectives that, when attained, signify completion
   (c) A temporary endeavor undertaken to create a service
   (d) Comprises of several interrelated projects managed in a coordinated way

3. Which of the following refers to management of similar but unrelated projects?
   (a) Project management
   (b) Program management
   (c) Portfolio management
   (d) Operations management

4. Which of the following is NOT a triple constraint for projects?
   (a) Time
   (b) Cost
   (c) Risk
   (d) Scope

5. Project sponsors have the greatest influence on the project during:
   (a) Planning
   (b) Execution
   (c) Control
   (d) Termination

6. Complex projects, involving cross-disciplinary efforts, are most effectively managed by:
   (a) Multiple project managers
   (b) A functional organization
   (c) A strong matrix organization
   (d) A strong traditional manager

7. What type of project organization does the project manager have total authority?
   (a) Functional
   (b) Strong matrix
   (c) Balanced matrix
   (d) Projectized

8. Which of the following organization structures have project managers reporting to a manager of project managers?
   (a) Functional Organization
   (b) Balanced Matrix Organization
   (c) Strong Matrix Organization
   (d) Traditional Organization

9. Which of the following organization structures cannot be used to create a Composite Organization?
   (a) Functional Organization
   (b) Balanced Matrix Organization
   (c) Strong Matrix Organization
   (d) Projectized Organization

10. Which of the following curves can be used to estimate the cumulative development effort over time?
    (a) Rayleigh curve
    (b) S-curve
    (c) Weibull distribution
    (d) Exponential curve

References

1. Project Management Institute. *A Guide to Project Management Body of Knowledge* (PMBOK), 5th Edn. PMI, 2013.
2. BS6079-1:2010 Project Management. *Principles and Guidelines for the Management of Projects*. British Standards Institute, 2010.
3. PRINCE2 Electronic On-Line Manual. Central Computer and Telecommunications Agency (CCTA).
4. Juran, J. M. *Juran on Planning for Quality*. New York: The Free Press, 1988.
5. Project Management Institute. *A Guide to Project Management Body of Knowledge* (PMBOK), 5th Edn. PMI, 2013, p. 50.
6. Project Management Institute. *A Guide to Project Management Body of Knowledge* (PMBOK), 5th Edn. PMI, 2013, p. 40.
7. *Crosstalk Magazine*. December 2000.
8. Project Management Institute. *Project Management Institute Code of Ethics and Professional Conduct*. PA, USA: PMI, Inc., 2006.
9. Professional Engineers Ontario. *Professional Engineers Ontario Code of Ethics*. Canada: Professional Engineers Ontario, 1990.

CHAPTER THREE

Key Project Management Artifacts

There are many project management artifacts in the process of managing projects. Of course, some are more important than others. Here, we present the four mandatory artifacts that MUST be included. Others can be skipped in the name of simplicity if the project is small. But the following four have to be present for any project, and we discuss them now. You will learn what they are, why they are important, and we will use a template to direct the creation of each. You may not be familiar with some of the terminologies in the templates, but more will be explained in subsequent chapters.

- Project Charter — sanctions a project.
- Project Management Plan (PMP) — the totality of all plans for the overall project.
- Integrated Change Control Plan (ICCP) — ensures the discipline of preparing, accepting and controlling changes to a system baseline.
- Quality Assurance Plan (QAP) — plans for quality goals, quality assurance and quality control of the project.

3.1. Project Charter

A Project Charter is typically issued by the project initiator or project sponsor, although it may be prepared by the Project Manager (PM) or team leader. The Charter sanctions the project via formal signature or electronic approval of the stakeholders or Project

Board. The Project Initial Meeting in PRINCE marks the official start of the project. (The Project Board approves the plan for the Initiation Stage at this meeting.)

Amongst other things, the Charter details the rights and obligations of the PM. For example, the PM must ethically apply himself to the best of his ability to manage the project in a professional way. The Charter must detail the PM's right to change the Project Plan (especially cost and timelines) if the scope or requirements are changed (more on that process later in Chapter 5). It should detail the level of quality expected.

The Charter is like a contract. When in doubt, READ THE CHARTER. Of course, the Charter should have a Management Reconciliation plan clearly enunciated. We cannot stress enough how important this is for the PM. If it is not stated, then the boss calls all the shots by default. The approach is similar to a Marriage Contract, where you plan your divorce and division of goods when you are full of love for each other. If you never need it, fine. But if you do, there is a logical, reasonable, non-hateful procedure to follow. Lawyers may not like it, but you will. In the same fashion, there is a reconciliation procedure to follow in the Charter. It should be simple and fast. Either side can invoke the procedure. You explain verbally your case to a neutral party who has the knowledge to understand the issues. The other side does the same. Then the arbitrator has an hour, for example, to make a binding decision. Fair and Fast. Most importantly, the project work must not be delayed.

The Charter should also contain the Vision and Mission Statements of the project and show how they align with those of the organization. The business case must also be included. Then the objectives of the project need to be clearly laid out. Finally, very high-level estimates of costs and time-to-complete need to be given. If there is to be any knowledge transfer as a result of doing the project, it must also be listed.

The organization should have both a Vision Statement and a Mission Statement. A Vision Statement describes the desired state of the organization's future. Where will it be in a decade, for example, and what will we look like then?

The Vision Statement of the organization must answer the questions:

1. What do we want to look like in *n* years?
2. What is "*n*"?
3. How can we transition from now to then?

The Mission Statement of the organization describes where we are today and is the fabric against which the goals and objectives of the organization can be compared. Normally, it is quite concise. It answers the fundamental questions of:

1. What does the organization do?
2. For whom does it do it?
3. How does it do it?

For example, a university might have a Mission Statement like "we aim to do world-class researching and give world-class teaching". Not good! Using terms like "world-class" is unacceptable puffery. Worse, there are two goals here. What if the budget were cut 20%? What would suffer? Research or Teaching or both?

The point of defining both the Vision and Mission Statements is that you incorporate them in your Project Charter, clearly showing how the benefits of the project will aid in achieving these stated goals.

Vision and Mission Statements are not metricized. They are "warm and fuzzy" views of how we see ourselves today and in the future. But we need to refine these non-functional visions into measurable entities. We call these objectives. Objectives, to quote the PMBOK definition, "are things toward which work is to be directed, a strategic position to be attained, a purpose to be achieved, a result to be obtained, a product to be produced or a service to be performed"[1]. Objectives are clear, concise, and can be measured. For example, the use of the term "world-class" could be replaced with "must rank at least Number 5 in the Maclean's Yearly Ranking of Engineering Schools". Following Galileo, if we do not have metrics to measure our objectives, we must make the unmeasurable, measurable. Thus Vision and Mission Statements give rise to specific objectives that show whether we are achieving our goals. The Project Charter should also have Project Objectives that are aligned with the corporate objectives.

Objectives should be SMART, that is:

- Specific
- Measurable
- Agreed to
- Realistic
- Time-constrained

The statement must be clear, concise, and specific. For example, "the response time to any inquiry must be less than 1 second". There must be a metric associated with the objective. It could be binary (met or not met). Normally, it will be a real number such as seconds or kilograms. The objectives must align with the Vision and Mission Statements of the organization and be agreed upon by management and the project manager. The objective must be realistic and do-able by the team in a reasonable amount of time. (Fred must run a mile in 3 minutes is not a reasonable objective.) Finally the objectives must be time-constrained. It must be completed within a reasonable period of time.

Other desirable properties beyond SMART include:

- verifiable
- aligned
- couched as a deliverable
- single-purposed

Of course, we will have many objectives (to start with, a cost objective, a time objective, and a quality objective). Remember that we can satisfy any two of those objectives but not all three. That means we will have to rank the objectives. Which is the most critical; which is the second most important, and so on. We can place the objectives in a matrix, with various stakeholders listed along the top and the objectives along the side. Each stakeholder will rank the objectives according to her view. Of course, the rankings will vary. Imagine how sales would rank the objectives compared to engineering! It is part of the responsibility of the PM to know these rankings and to explain why a certain ranking was chosen for the project. Careful stakeholder management will pay handsome dividends during the project execution.

Here is a template that you can use for preparing the Charter.

1. General requirements
2. Business alignment (with the Vision and Mission Statements of the organization)
3. Project purpose and objectives
4. Assigned PM and authority
5. Summary milestones
6. Summary budget (estimating the inflow and outflow of funds to the project)
7. Stakeholders
8. Functional organizations and their influence
9. Measurable objectives and success criteria
10. Constraints and assumptions (each one of which will turn into a project risk)
11. Business case
12. The PM's rights and obligations (his authority level)

Let us apply this to SPENTFUEL as an illustration of the use of the template.

3.1.1. The charter of SPENTFUEL

1. *General requirements*

The general purpose of SPENTFUEL is to build a database that will contain the location of all of the spent nuclear fuel cells in North America, with potential expansion to cover these sites globally. The database will be accessible to all authorized sites in the catchment area. Using a defined common template, each site will populate its portion of the database with its local records.

The template will contain at least the following information:

1) location and responsible company/agency
2) burial date
3) amount buried
4) nature of the burial details
5) regulatory authorities associated with burials

The software application must be portable and hardware/operating system independent. It must be of very high quality and be capable of easy and cost-effective maintenance. All communications will use the TCP/IP protocol suite.

2. *Business alignment (with the vision and mission statements of the organization)*

The mission statement of Vacux is to firstly provide, to government agencies, software of exceptional quality and secondly, at a reasonable price. This project aligns perfectly with our mission statement. We are well qualified to do the work and have extensive experience in constructing applications that will perform well in safety-critical areas.

3. *Project purpose and objectives*

The primary purpose of the project is to construct software that will meet the requirements stated above and do so at an acceptable cost. Specific project objectives include:

1) build the software at contracted cost,
2) build the software on contracted schedule,
3) produce software of exceptional quality,
4) produce software that is almost error-free,

5) construct software that is user-friendly, and
6) construct software that is platform-independent.
   See 9 below.

4. *Assigned PM and authority*

The project manager will have discretional control over the assigned budget.

5. *Summary milestones*

- Start + 3 months: requirements sign-off
- Start + 6 months: detailed design sign-off
- Start + 7 months: PMP sign-off
- Start + 12 months: prototype sign-off
- Start + 15 months: local acceptance test
- Start + 18 months: complete delivery of product, documentation, etc.

6. *Summary budget*

• Start:	$30,000
• Start + 3 months: requirements sign-off	$60,000
• Start + 6 months: detailed design sign-off	$20,000
• Start + 7 months: PMP sign-off	$95,000
• Start + 12 months: prototype sign-off	$30,000
• Start + 15 months: local acceptance test	$30,000
• Start + 18 months: complete delivery of product	$45,000

7. *Stakeholders*

7.1 Vacux BoD, managers, team members, Project Manager
7.2 DND ADM
7.3 CNA ADM
7.4 Citizen Advisory Committee
7.5 Parliament Advisory Committee

8. *Functional organizations and their influence*

8.1 *Vacux and its software unit*
We will populate our project team from this functional unit. The PM will also be chosen from this unit. In addition, expertise on different software platforms will be provided here.

8.2 *Vacux's database professionals*
This unit will provide database consulting for our DB design.

8.3 *Vacux's data communications experts*
These people have TCP/IP communications experience and will advise our communications people.

9. *Measurable objectives and success criteria*

To prove that our objectives have been met, we propose the following metrics:

- Objective 1. The CPI (defined in Chapter 7) must be within 5% of 1.0.
- Objective 2. The SPI (defined in Chapter 6) must be within 5% of 1.0.
- Objective 3. The McCall Index (defined in Chapter 8) must be above 8.
- Objective 4. The defect rate must be above 5 Sigma in terms of Lines of Code (less than 1 defect per 100,000 lines of code).
- Objective 5. User trial tests must be stress-free for 1 hour of use.
- Objective 6. Vacux will demonstrate program execution on the following platforms:
  - Windows 7 32-bit and 64-bit versions
  - Linux 3.5
  - Snow Leopard 9.1
- Objective 7. The application will be written in Java version 7.1 with no external references.

10. *Constraints and assumptions*

10.1 Money will be available at the appointed milestones.
10.2 Appropriate software skill-sets are available for the team selection.
10.3 Members of the functional units are available when needed.
10.4 Software and hardware platforms will be available when needed.
10.5 If training is necessary, it will be provided and the schedule adjusted accordingly.

11. *Business case*

The organization has constructed similar projects in the past and has brought the projects in on-time and on-budget, to the satisfaction of the clients. All of our team members are available for this project and can be assigned for duty now. We anticipate making a reasonable profit on the project.

12. *The PM's rights and obligations*

12.1 The PM has the right to change the PMP if the scope of the project is changed after sign-off.
12.2 If the PM and the manager (or equivalent) disagree and an accommodation cannot be made, each side will write up their relevant concerns and give the reports to the VP of Operations. That person will decide within 2 hours, the resolution of the problem.
12.3 The PM has the right to allocate resources to the project, subject to company policies.
12.4 The PM has the obligation to exercise a Duty of Care in the running of the project, following the procedures of the organization.

This is a skeleton of the Charter and you may want to add additional points.

3.2. Project Management Plan (PMP)

The development of the PMP is a key success factor for any project. The PMP defines, integrates and co-ordinates all project management subsidiary plans (i.e. scope, requirements, time, cost, quality, human resource, communications, risk and procurement, in accordance with the PMBOK) into one document. In particular, the PMP for a software project is known as the Software Project Management Plan (SPMP). Another component that may be part of the PMP or split into a separate document is the ICCP. The QAP may be a separate document too. In fact, there will be up to 15 subplans built into the final PMP. This suggests that we should be using a relational database to store them and permit extractions of each of the subplans when needed. Possible subplans include the following (with the relevant PMBOK references):

1) Scope management plan (5.1.3.1)
2) Schedule management plan (6.1.3.1)
3) Cost management plan (7.1.3.1)
4) Quality management plan (8.1.3.1)
5) Process improvement plan (8.1.3.2)
6) Human Resource management plan (9.1.3.1)
7) Communications management plan (10.1.3.1)
8) Risk management plan (11.1.3.1)
9) Procurement management plan (12.1.3.1)
10) Milestone list (6.2.3.3)
11) Resource calendar (6.4.1.4)
12) Schedule baseline (6.6.3.1)
13) Cost baseline (7.3.3.1)
14) Risk register (11.2.3.1)
15) Stakeholder management plan (13.2.2.1)

The Project Plan is the summation of all of these subplans. In Appendix B, we include a sample of the IEEE Software Project Management Plan (IEEE standard 1058 1–1987) [2] for SPENTFUEL.

Now the word "software" is unnecessary and we shall refer to it as the PMP. Here is the PMP template:

0. General
- Title page
- Signature page
- Change history
- Preface
- TOC
- List of Figures
- List of Tables

1. Overview
- 1.1 Project Summary
  - 1.1.1 Purpose, Scope and Objectives
  - 1.1.2 Assumptions and Constraints
  - 1.1.3 Project Deliverables
  - 1.1.4 Schedule and Budget Summary
- 1.2 Evolution of the Plan
- 1.3 Charter

2. References

3. Definitions

4. Project Organization
- 4.1 External Interfaces
- 4.2 Internal Structure
- 4.3 Roles and Responsibilities

5. Managerial Process Plans
- 5.1 Start-up Plan
  - 5.1.1 Estimation Plan
  - 5.1.2 Staffing Plan
  - 5.1.3 Resource Allocation Plan
  - 5.1.4 Project Staff Training Plan
- 5.2 Work Plan
  - 5.2.1 Work Activities
  - 5.2.2 Schedule Allocation

5.2.3 Resource Allocation
5.2.4 Budget Allocation
5.3 Control Plan
5.3.1 Requirements Control Plan
5.3.2 Schedule Control Plan
5.3.3 Budget Control Plan
5.3.4 Quality Control Plan
5.3.5 Reporting Plan
5.3.6 Metrics Collection Plan
5.4 Risk Management Plan
5.5 Communications Management Plan
5.6 Closeout Plan

6. Technical Process Plans
6.1 Process Model
6.2 Methods, Tools and Techniques
6.3 Infrastructure Plan
6.4 Product Acceptance Plan

7. Supporting Process Plans
7.1 Configuration Management Plan
7.2 Verification & Validation Plan
7.3 Documentation Plan
7.4 Quality Assurance Plan
7.5 Reviews and Audits
7.6 Problem Resolution Plan
7.7 Subcontractor Management Plan
7.8 Process Improvement Plan

8. Additional Plans
8.1 Safety
8.2 Privacy
8.3 Security
8.4 Installation/rollout
8.5 Maintenance
8.6 Additional subplans

Annexes
Index

One may ask how big a PMP document is. Well, it depends on the nature, size, and complexity of the project, as well as the capability and maturity of the organization. We have seen PMPs from engineering that vary from 10 to 20 pages, to over 1,000 pages. Different organizations may have standard templates for PMPs or even references (within the PMPs) to standard operating procedures of the organization, thus minimizing the duplication of content and overall size of the PMPs. The key to remember is that the PMP must be a living and working document. The PMP is progressively being detailed; parts are exposed to great detail while other sections to be done later are sketched out. This is also called rolling-wave planning. Any new hires to the project can easily join the existing team by reading and understanding the PMP. The team should also make reference to the PMP for any doubts while working on the project.

3.3. Integrated Change Control Plan (ICCP)

The ICCP ensures the discipline of preparing, accepting, and controlling changes to a system baseline. Changes to be managed include additions, corrections, modifications, and enhancements that are applicable equally to hardware, software, and documentation so that all change aspects are known and documented. The user would typically specify the need for configuration management as a requirement. Such a requirement is required for all project documents and artifacts regardless of their complexity, as any artifact is subject to change during a project. Note that Change Control is more inclusive than the normal Configuration Management Plan (CMP). We make reference to PRINCE that says, "nothing moves, nothing changes, without authorization" [3]. CMP normally identifies, controls, keeps track of, and verifies a specific list of tangible items for the project's products such as car parts or modules in a complex software package. ICC manages all aspects of the project, not just parts lists.

Some entities requiring change control management include:

- Requirements
- Software libraries

- Project management tools
- Documents
- Media and interfaces of the test and delivered systems

The ICCP is concerned with the formal control of changes. When changes are made, ICCP must determine that they were made appropriately and that anyone needing to know about the change is properly informed and all references to the pre-change version are hidden. Some of ICCP activities include:

- Identifying that a change has occurred or needs to occur.
- Making sure you have a procedure in place so that only officially approved Change Requests (CRs) occur.
- Reviewing and approving (or rejecting) the CRs.
- Managing the approved changes when they occur by regulating the flow of requested changes.
- Maintaining the integrity of established baselines (no sneak releases!).
- Reviewing and approving all recommended corrective and preventive actions.
- Controlling and updating relevant documents such as Project Scope Statement, Project Management Plan, Risk Plan, and so on.
- Documenting the impact of requested changes.
- Validating defect repair.
- Controlling project quality to standards (and auditing) based on quality reports.

The ICCP should also identify the process for version control and auditing of such process.

Now let us take a closer look at the problem. When an item is placed under ICC, it is baselined. That is, it has formally been accepted as a delivered item. It can still be changed but to do so, you must follow a strict process. Imagine you are sitting in a restaurant ordering food with your friend; the waitress asks you what you would like to eat and you say "the steak" and the waitress writes that on her pad. As she turns away, you suddenly remember that you had a big lunch and steak would be too much. So you ask the waitress to replace it with the ham and shrimp. She crosses out "steak" and adds "ham and shrimp". As she heads into the kitchen, you suddenly remember that your religion forbids you to eat meat on Friday and rush to the kitchen door and shout "make that fish please". She crosses out the old and replaces it with fish and hands the order to the chef. The chef takes a fish and chops its head off, guts it and places the carcass in a frying pan. Your order is now baselined. You could, of course, order something else but you would then be getting two meals. Until an item is baselined, it can be changed at will. Once it is baselined, we have to go through a very formal, written procedure to change the baselined item, as we will describe.

A clever way of describing the execution of the project plan is to view it as a constant thrust to baseline all project items. How is an item baselined? In the PMP, you will describe all of the activities that must be done for project completion. Each item (we call them work packages, see Chapter 5) has a deliverable; a tangible object that is created as a project artifact. That deliverable will have acceptance criteria which must be met before it can be baselined. In our restaurant example, there must be a fish there before it can be baselined. Again, a person outside of the project team must deem that the deliverable has met the acceptance criteria and can be baselined. Then it is entered into the ICC Database.

The ICC Database is a repository for baselined items. It can be primitive (e.g., 3 × 5 inch white cards) or it could be a massive Oracle database.

Now we look more closely at the ICCP process. First, we must identify the objects needing ICC. The smallest item under ICC is called the Configuration Item or CI. A CI is a named controlled object.

It can be a subplan, a subsection of that plan, or even a subcomponent of that subsection. The decision of how fine-grained to make the CIs depends on the nature of the object to be controlled. Like many other aspects of Project Management, we must trade-off the extra cost of describing CIs against the inconvenience of having to lock out a huge CI object. For example, pieces of code are CIs. If we clump up thousands of lines of an application into a single CI, then we would have to take out the entire CI from the database when modifications were needed, thus making it unavailable to anyone else. If we carved up the application into objects or procedures and turned those into CIs, then we could have multiple teams doing concurrent maintenance on many of those CIs at the same time. If we go to the other extreme and make each line a CI, then we have enormous overhead describing the CI; much greater than just writing the line in the first place.

3.3.1. The ICCP for SPENTFUEL

A minimum CI list for SPENTFUEL might include the following:

1. System Specification
2. PMP
3. Requirements Specification
   (a) executable or "paper" prototype
4. Preliminary Users Manual
5. Design Specification
   (a) data design document
   (b) architectural design document
   (c) module design document
   (d) interface design document
   (e) object descriptions
6. Source Code Listing
7. Testing
   (a) procedures
   (b) test cases and recorded results
8. Operational Manuals
9. Executable Modules
   (a) modules of executable code
   (b) linked modules

10. Database Description
    (a) schema and file structure
    (b) initial content
11. As-Built User Manuals
12. Maintenance Documents
    (a) software problem reports
    (b) maintenance requests
    (c) Engineering Change Orders (ECOs)
13. Standards and Procedures

Each ICC CI must be described. A sample template might be as follows:

1) Name of CI
   - Version Number
   - Object Class of CI
   - Status of this CI (locked or unlocked)
   - If locked, owner of lock

2) Description of CI's Purpose
3) List of "resources" that it needs to function
4) A "realization" of the CI
5) Name of creator
6) Date of initial creation
7) Date of this instantiation
8) Reason for this instantiation
9) Pointer to test cases
10) Party to sign-off this instantiation

Thus each CI would be listed in the ICC Database, with the identification header and the "instantiation" of the text. That is a vague term to denote that the CI could be in various forms. It might be a set of code lines; it might be a section of a manual in text, it might be a specific version of a Microsoft Operating System, The totality of the ICC Database at the end of the project would be all of the artifacts that had been both produced by the Project Team and signed-off by the appropriate authorities.

3.3.2. Change control process

Once an object has been entered into the ICC Database, it has been baselined. If there needs to be a change, a formal change procedure must be followed (that process is also a configuration item — CI!). Changes are necessary, but they must be done in an orderly manner. For a PM, to reject all changes would be a guarantee of building a product that no one wants to use. But a free-for-all change process would lead to chaos and the destruction of the project.

Let us track a change. Someone identifies the need for a change of a CI. That must be written up as a Change Request (CR). That "someone" could be anyone associated with the project. Then the CR is vetted by the PM to be sure the change is not frivolous. The CR is then passed on to a person who can draw up a mini-project plan to implement the change, including costs, schedule, and benefits to the project. Again, the PM will verify the calculations. The PM then takes the CR to the Change Control Board (CCB). The CCB is a committee consisting of the manager, the PM, stakeholders selected for their expertise, the client and anyone else management thinks can help in the decision making. Note that changes are almost always costly in both time and money. So the benefits of the change MUST outweigh the additional costs and time accretions. The CCB will then decide. It, of course, could call in additional experts to help in that process. Once decided, either the change is denied and the author informed, or the change is accepted. Now the PM has to implement the change. It can either be a single action or a true project plan that has to be integrated into the original plan. Either way, it receives an Engineering Change Order (ECO) number. That tracking number is used to follow the progress of the change. This is yet another process that must be tracked and improved upon.

When the change is begun, the leader of the team doing the ECO must check out all modules related to the change.

"Check-out" means that no one else can modify these CIs until they are reinserted into the database and unlocked. The change is made and all testing related to these modifications are redone. In testing parlance, this is called a "regression test". We want to be sure our modifications do not break other components. Once that change is completed and "signed-off", the CIs are then reinserted into the Database and unlocked.

3.3.3. ICCP summary

ICCP is vital to the health of a complex process. Only by enforcing a strict ICC discipline upon the project team can we guarantee the integrity of our deliverables.

We conclude by giving an example of ICC in real life. A couple of decades ago, one of us was on sabbatical in Denmark and was told the following anecdote. After the debacle of the Second World War, in which Denmark was captured by the Germans in a day, the Danish military resolved to always be ready for any future invasion. Thus they introduced conscription for all Danish males. All must spend a year in the military so they can be recalled in the event of invasion. My colleague Sven, a formal Computer Scientist, was swept up into this system. What to do with a theoretical Computer Scientist, the Army pondered? Military structures have paper manuals for all sorts of artifacts: guns, tanks, aircraft, etc., and like all manuals, these need to be updated periodically. Enter Sven and his ICCP! He spent his year developing an algorithm for manual updating in the field under enemy fire. He solved the following problem; a soldier is updating a manual in the midst of a battle when alas, he is killed. Another soldier immediately rushes up and must determine how much of the updating has been done and where he should begin to continue the work so that the manual is always in a consistent state. We have ever since the image of the Danish Army having the most current manuals imaginable but always losing the war because the soldiers are so busy with their updates instead of shooting at the enemy! In conclusion, we note that ICC is incredibly important. Sven's solution, by the way, is non-trivial.

3.4. Quality Assurance Plan (QAP)

The QAP specifies the quality standards that are relevant to the project, activities to ensure meeting of the quality standards and ways to correct deviation from the standards. It also specifies how, when and by whom each product will be tested against its quality criteria. Verification and validation will typically be done by an independent party. This is known as Independent Verification and Validation (IV&V); see Chapter 8. Some entities for the QAP are:

1) quality baseline (metrics from previous projects),
2) quality metrics for current project (based on quality goals and improvement plan from the quality baseline),
3) checklists,
4) processes and procedures for audit (to assure quality), and
5) processes and procedures to identify and correct deviation from quality metrics.

In assessing quality, the most important step is to determine the level of quality needed. Increased quality means increased expense. We need to define the level of "good-enough" quality. Let us give a simple example. Another Danish story. On that sabbatical, one of us had rented a house with a hot water heating system. The main radiator in the house had its control, a simple lever, snapped off. Being an engineer, he immediately rushed down to the hardware store to buy a little wrench, the type you can easily buy for a dollar. Of course, it would snap if used in a serious application but it would do for the radiator. Imagine his surprise when the only wrench for sale was a vanadium beauty from Germany that was priced at $50. The quality far surpassed the task at hand.

So first thing we do, determine the level of quality we need.

Here is the IEEE 882–1986 [4] QAP template as an example.

1) QA Organization and Resources
    - Organization structure
    - Personnel skill level and qualifications
    - Resources

2) QA standards, procedures, policies, and guidelines
3) QA documentation requirements
4) QA software requirements
5) Evaluation of storage, handling, and delivery
6) Reviews and audits
7) Configuration requirements
8) Problem reporting and corrective action
9) Evaluation of test procedures
10) Tools, techniques, and technologies
11) Quality control of contractors, vendors, suppliers
12) Additional control
13) Miscellaneous control procedures
14) Project specific control
15) QA reporting, records and documentation
16) Status reporting procedures
17) Maintenance
18) Storage and security
19) Retention period

3.5. Chapter Summary

We conclude by repeating that these four artifacts are vital for the success of any project. By getting these right, your chances of success are greatly improved. The four artifacts, while not sufficient, are certainly necessary for project success.

3.6. Multiple Choice Questions

1. Which of the following is NOT a part of the Charter?
   (a) Requirements
   (b) Formal recognition of the existence of the project
   (c) Definition of project scope and objectives
   (d) Definition of authority of PM

2. Configuration Management is:
   (a) Used to ensure that the description of the project product is correct and complete
   (b) The creation of the work breakdown structure
   (c) The set of procedures developed to assure that project design criteria are met
   (d) A mechanism to track budget and schedule variances

3. Which document provides the information of whom has approval authority for revisions in scope of a project?
   (a) Resource assignment matrix
   (b) Change control plan
   (c) Project charter
   (d) Organization chart

4. Sign-off is important because:
   (a) The party to sign-off wants the glory
   (b) The PM gets another part of the project off her back
   (c) The ISO guidelines say so
   (d) Because it proves that part has been successfully done

5. The Prime Directive of the Project Manager is to:
   (a) Bring the project in, on time, on budget, on scope
   (b) Help team members become better at project management
   (c) Advance the careers of team members
   (d) Keep the stakeholders satisfied

6. When a project falls seriously behind schedule, the most appropriate response is to:
   (a) Fire the project manager
   (b) Make the team work overtime
   (c) Add more people to the project
   (d) Revisit the project plan

7. "Gating" means:
   (a) You are on the way to becoming as rich as Bill Gates
   (b) Keeping costs under control as in controlling cows leaving a field
   (c) Team members are rotating through the team at a great rate
   (d) Justifying funding to proceed to the next stage of the project

8. What does the I stand for in IV&V?
   (a) Independent
   (b) Individual
   (c) Integrated
   (d) Indivisible

9. Projects can often be optimized for Shortest Schedule, Least Cost, and Best Quality:
   (a) Simultaneously, all three can be achieved by careful scheduling and budget control
   (b) So that any two of the above can be achieved at a given time, but not all three
   (c) We are not as concerned with optimization, as we are with balance
   (d) Only one can be achieved

10. Configuration Management is a technique to:
    (a) Perform overall change control
    (b) Execute project plan
    (c) Collect requirements
    (d) Perform quantitative risk analysis

3.7. Exercises

1. From the Project Manager's point-of-view, what are the 3 most important things that must be in the Charter and why?

2. Set a SMART goal for yourself.
3. For undergraduate students, write a Charter for your 4th year design thesis/project.
4. For graduate students, write a Charter for your Masters or PhD thesis.
5. You are the Project Manager on a small project that is estimated to take 6 months. Your organization has no formal ICCP in place. You must design your own. With the help of 3x5 inch index cards, what would you do?

References

1. Project Management Institute. *A Guide to Project Management Body of Knowledge* (PMBOK), 5th Edn. PMI, 2013, p. 548.
2. IEEE Standard 1058 1–1987. Software Project Management Plan. IEEE, 1987.
3. PRINCE2 Electronic On-Line Manual. Central Computer and Telecommunications Agency (CCTA).
4. IEEE Standard 882–1986. Quality Assurance Plan. IEEE, 1986.

CHAPTER FOUR

Engineering Economics and Project Management

4.1. Introduction to Engineering Economics

Benjamin Franklin said it best; "time is money" [1]. This is of the greatest importance to the engineering project manager. The manager must justify the cost of the project in terms of return on the investment. In this chapter, we examine several well-known ways of doing this.

4.2. Time Value of Money and the MARR

4.2.1. Simple interest

Suppose that I have $100 in my pocket. I have no need to spend it at this moment. What should I do with it? I could buy 1,000 jelly-beans at a dime a piece, but I am not hungry right now. I could keep it in my pocket or put it in the freezer for safe-keeping anticipating the day when a jelly-bean crave overtakes me. Suppose that I left that $100 bill there for a year. What would it be worth? Of course, it is denominated at $100. But it cannot purchase 1,000 jelly-beans a year from now. Why not? Because of inflation. Inflation is the measure of an aggregate value of normal products through a specified time period. Sometimes and in some places, inflation can run amuck. But in Canada, our inflation rate has averaged about 2% per year. Thus postponing my jelly-bean purchase means that I can only buy 2% less or 980 jelly-beans, for that 100 dollar bill. The value of my money has been eroded by inflation. Inflation is like rust; constantly eating away at money that is not keeping pace with inflation. Why is that? Because the money is not "working" for us. How can money "work"? By paying us back an interest on that money. Suppose I deposit my \$100 in a bank savings account that promises me 3% interest. In a year, I will have \$103 in my account and can buy more than 1,000 jelly-beans a year from now. Note that inflation also eats away at my investment rate and my REAL rate of return is only 1%.

4.2.2. Project timeline, cost, and benefit

Why do we do projects? For most engineering cases, we do them to produce a product that we can then sell at a profit. In other words, we will only work if there is a strong chance that we can make more money than the project costs. Or in other terms, we would not get out of bed in the morning, unless there was an economic incentive to do so. As we stated at the beginning, money is time dependent. Under normal conditions, money decays over time and we need to factor this into long-term decision making.

To illustrate this, we use a simple time-line as shown in Figure 4.1(a). Time 0 is time-present. Normally we mark the time units in terms of years but any fixed time period could be used. Figure 4.1(a) is a time-line for a 5-year project.

The purpose of the time-line is to show the flow of money into and out of the project. By convention, we assume the cost of the project occurs at time zero and is represented by a downward arrow. Positive money flows (or profits) are indicated by upward flows, at the end of the time period. Again this is convention. In reality, money would flow in spurts all through the time period. Similarly, we assume that there is an interest rate that we assume for the length of the project (see next section). That interest rate will

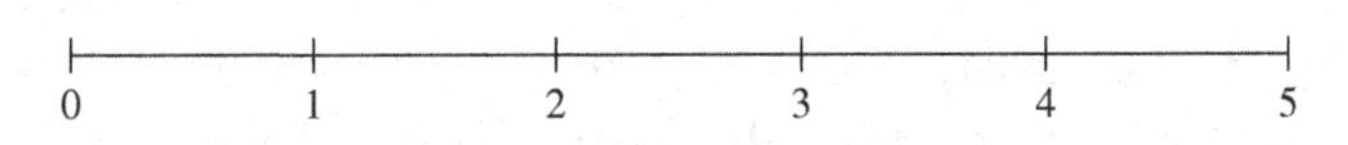

Figure 4.1(a). Timeline for a 5-year project.

Figure 4.1(b). Cost and benefits for a 5-year project.

be the same for the length of the project, again a very simplifying assumption.

Suppose that our Project A costs $10,000 to complete. We call the initial investment the "First-Cost" and of course, it is negative. We start selling the product and it nets $3,000 for each year for 5 years. Then our time-line would look like Figure 4.1(b).

4.2.3. The MARR

What should a project "make" in terms of an investment? In Project Management, we call this the ROI — Return On Investment, and this is the most important metric for project acceptance. Our ROI here is simple: $10,000 out and $15,000 in. We have made a nice profit of $5,000. But the value of money decreases with time. In Economics, this is called the time-value of money. That $3,000 a year from now is not worth as much as $3,000 today. Why not? Why does money change over time? Why is it not constant?

To see why this is so, consider the price of gasoline. When one of us first drove (1956), gasoline was 10 cents a liter. Today (2014) it is about $1.30. If we tracked the price of a packet of cigarettes, you would see an increase from 40 cents to $7.00 over the same time period. Now this could be explained by many things: demand, scarcity, taxes, etc. But in general, everything has increased in price. But then, so have salaries. One of us started an engineering job in 1963 for the princely annual salary of $5,400. Today an engineer begins her career for a salary around $65,000. What has happened is that over time, money has decreased in value. Our term for that is inflation. If we spend $1.00 today to buy a set of goods, we can expect to spend $1.00 times the inflation rate to buy the same set of goods in a year. If the inflation rate is 2%, then $1.02 would be needed to buy the same goods. In other words, the worth of our money has declined. Thus in making economic decisions, we have to factor in the effect of inflation. Moreover, we need to consider the "opportunity cost". Economically speaking, that is the money we have lost by NOT doing something. For example, by attending university for one academic term, you give up the chance of making $10 an hour for 40 hours times 32 weeks or $12,800. By not flipping burgers at a fast food place, your opportunity cost is $12,800.

If we had not invested our $10,000 in our project but had invested the $10,000 in a bank, what would it have become? Today the interest rate on Canada Savings Bonds (CSB, the canonical example of a safe investment) is about 1.5%. Our annual inflation rate has been relatively steady at 2%. Thus the investment has to make at least 3.5% to beat the CSB investment and inflation. But all investments are risky. If there is no risk involved then 3.5% may be acceptable. But what if there is significant risk? Then we would demand a much higher rate of return. Do not forget about profit (that nasty word to socialists), which is what motivates us to try new things. Thus we need a number that includes all of these factors: inflation, CSB yield, risk, and profit. That number we term the Minimum Attractive Rate of Return (MARR). The setting of the MARR is done by the Organization and not by the Project Manager. Note that MARR is also called the discount rate or even the interest rate. We use MARR as meaning all three.

Now we demand that any project accepted for funding must normally have an ROI > MARR. If not, it will not generate enough money to offset the risk and cost of doing the project. Note the use of the word "normally". Sometimes, we simply have to do projects.

If a law is passed by Parliament, that software project HAS to be done and the ROI is irrelevant. Even in the private sector, we may be running a project at a loss; to exclude a competitor from entering our field of competition, to keep a team together for a downstream successful project, etc. But for most cases, the ROI must be greater than the MARR.

4.3. Capturing the Time Value of Money

We need a mechanism for displaying the Time Value of Money. Let us begin with some simple definitions and abstractions.

- P = present sum today (time 0)
- F = future sum of money after the last of n time periods
- n = number of time periods (normally years)
- i = interest rate (normally percentage/year)
- A = annual cash flow per year

In our Project A example in Section 4.2.2, P = \$10,000, F = the total amount earned after 5 years = \$15,000, n = 5, i = 0%, and A = \$3,000.

4.3.1. Simple interest

Simple Interest is an amount of money expressed as a percentage of the original sum. If I were to borrow \$1,000 from my father at 5% interest, the interest per year would be \$1,000 × 0.05 or \$50. If the loan remained unpaid, that rate would remain constant. In other words, there is no "interest" on the interest. Bonds often pay out a fixed amount per time period based on the set interest rate. We can express this mathematically by writing:

$$F = P + P(n)(i) = P(1 + ni) \text{ for } n \text{ periods of simple interest } i.$$

Thus if I were to buy a 20-year bond for \$1,000 that yielded 5% per year, I would receive 20 × \$50 or \$1,000 interest over the life of the bond (i.e., 20 years).

4.3.2. Compound interest

With compound interest, we calculate the interest on the accumulated amount. For the first year, simple interest and compound interest are the same. But the second year, we calculate 5% on $1,050, not $1,000. This does not sound like much, but over time, the amount can grow exponentially. Here is a simple example.

Consider $25,000 invested at 10% compounded in Table 4.1.

Table 4.1. Compound interest table.

Year n	Total at start of year n	Interest accumulated at end of year n	Amount accumulated at end of year n
1	$25,000	$2,500	$27,500
2	$27,500	$2,750	$30,250
3	$30,250	$3,025	$33,275
4	$33,275	$3,327.50	$36,602.50

With simple interest (and putting our annual interest payments under our mattress), we would only have had $35,000 at the end of year 4. That is slightly less than the compound value of $36,602.50. But consider the difference after 100 years. Simple interest would have yielded 25,000 + 2,500(100) = $275,000. But compound interest would have yielded $344,515,000 (plus the original $25,000!). This is often called the miracle of compound interest; slow to accumulate but what an eventual accumulation!

4.3.3. Nominal and effective interest rates

Interest rates can be expressed in different ways, depending on the way compounding is done. The *nominal interest rate* is the annual rate with no compounding. The effective interest rate is the annual rate with compounding taken into account. Consider a nominal rate of 10%. Suppose it was compounded twice yearly. Then there would be interest on the interest for the second compounding. If we deposited $100, then at 6 months, the bank would give back 5% or $5, leaving us with $105. At year's end, we would receive \$105 + \$105(0.05) = \$110.25. Thus the effective interest rate (call it i_e), thanks to compounding, is really 10.25%. We can generalize this by noting that the effective interest rate, with a compounding of r periods per year becomes:

$$[P(1 + i/r)^r - P]/P \text{ or } (1 + i/r)^r - 1$$

Compounding monthly would give us an effective interest rate of 10.46%. Compounding on a daily basis would give the formula $(1 + i/365)^{365} - 1$, or an i_e of 10.52%. As one may see, the value of the i_e increases as the compounding period, r, increases but the increases are smaller and smaller. We can ask what the limit of this is (that is, continuous compounding or r = infinity). As we have seen, $F = P(1 + i/r)^{rn}$. If we let r approach ∞, and let $x = i/r$, then we have:

$\text{Lim}(1 + x)^{(1/x)in}$ and we know from Calculus that the Lim $(1 + x)^{(1/x)} = e = 2.71828$.

Thus $F = P(e^{in})$ and the $i_e = e^i - 1$ for one year.

4.3.4. Inflation

If we assume that we are comparing different projects, the effect of inflation will be the same on all of the projects so we can ignore it. If it is a factor, then we need to discount the future cash flows by the following:

$$i_f = i + f + if, \text{ where } f \text{ is the inflation rate.}$$

The important thing to remember is to treat the flows the same; inflation in or inflation out.

4.3.5. Equivalence of the time value of money

We have shown that a sum of money in one time period will have the same "value" to a different sum in another time period with respect to a given interest rate. For example, a \$1,000 today is equivalent to $1,100 in a year from now at 10% or $909 a year ago. If inflation were running at 10% (all else being equal), then a liter of gas costing $1 today will cost $1.10 in 1 year, $1.21 in 2 years and $0.91 a year ago.

Thus our time-line connects the P and the F in a fixed manner, relative to the assumed MARR. We consider two cases in working out compound interest problems: the equivalence of a single payment and the equivalence of a series of equal payments. We have essentially 5 variables; P, F, I, n, and A.

Because this is common to all compound interest calculations, standard notation has been adopted to represent the various interest factors. This is called the ANSI Standard Notation for Compound Interest (ANSI is the American National Standards Institute). The notation considers two of the three cash flow symbols (F, P, or A), the interest rate, and the number of time periods. The general form is $(X/Y, i\%, n)$, where

- X represents what is unknown
- Y represents what is known
- i and n represent input parameters; can be known or unknown depending upon the problem

Example: (F/P, 6%, 20) is read as: Find F, given P when the interest rate is 6% and the number of time periods equals 20. In our problem formulations, the standard notation is often used in place of the closed-form equation or Excel function. In standard Engineering Economics texts, tables are provided for tabulations of common values for $i\%$ and n. Several examples are given in Appendix C.

Essentially, we have a set of formulae consisting of four variables, of which three are known. Now the formulae can be solved in three ways:

1. Tables
2. Closed form formulae
3. Excel function

Our relationships are as follows:

- Single Payment Compound or Future Worth FW $(F/P, i\%, n) = P(1 + i)^n$
- Single Payment Present Worth PW $(P/F, i\%, n) = F\,(1/(1 + i)^n)$
- Uniform Series Future Worth FW $(F/A, i\%, n) = A\,((1 + i)^n - 1)/i$
- Uniform Series Sinking Fund $(A/F, i\%, n) = F\,i/((1 + i)^n - 1)$
- Uniform Series Present Worth PW $(P/A, i\%, n) = A\,((1 + i)^n - 1)/(i(1 + i)^n)$
- Uniform Series Capital Recovery $(A/P, i\%, n) = P(i(1 + i)^n)/((1 + i)^n - 1)$

Let us give a simple example of each.

- Single Payment Compound or Future Worth (FW) $= P(F/P, i\%, n) = P(1 + i)^n$

This expresses the future worth F of the present sum P. Consider our example of what will \$25,000 grow to in 100 years at 10%. $F = P\,(F/P, 10\%, 100) = 25000\,(13780.6) = 344{,}515{,}000$. Where did that number 13,780.6 come from? It comes from the 10% table (Appendix C) at row 100 in the F/P column.

Alternatively, we can use our calculator and punch in $FW = P(1 + i)^n = 25000(1 + 0.1)^{100} = 25000(13780.621) = 344{,}515{,}308.50$

Again, we can use the EXCEL function FV.

The call is FV (rate,nper,pmt,pv,type) where

- rate is the MARR
- nper is number of periods
- pmt is payment per month
- pv is present value
- type = 0 if payment is at the end of the period, 1 if at the start

Therefore we would write FV(0.1,100,0,25000,0) and FV returns –344,515,308.50. Use ABS to remove the negative value.

Note that all three methods are equivalent. But beware of the false sense of accuracy that the calculator and EXCEL gives. Even four digits are excessive with the assumptions we make in our model!

- Single Payment Present Worth PW = $F(P/F, i\%, n) = F\ (1/(1+i)^n)$

  This expresses the amount I have to put aside today to achieve F in n years at $i\%$. If *I* want to buy a $100,000 car in 10 years, what must I deposit today into my savings account? Assume that it will yield 10% per annum. Again, I know F, i, and n. Thus we solve for P, reading down the P/F column of the 10% table to row 10.

$$PW = 100000(P/F, 10\%, 10) = 100000(0.3855) = 38{,}550$$

  We note that this is simply the inverse of the FV function.

  With the calculator, we enter $100000/(1 + 0.1)^{100} = 38{,}554$

  In EXCEL, we use the function PV (rate,nper,pmt,fv,type) with the same definitions as above.

  Thus PW = PV(0.1,10,0,100000,0) = −38,554.33. Again note the negative formulation (use ABS to remove).

- Uniform Series Future Worth FW = $A\ (F/A, i\%, n) = A\ ((1 + i)^n - 1)/i$

  Here we assume that we are making an annual payment of A dollars for a number of years. What will it be worth in n years (or payment periods)? Suppose an engineer deposits $1,000 per year into a retirement savings account for 30 years. What will her nest egg be then?

  We write this as $1000(F/A, 10\%, 30) = 1000(164.494) = 164{,}494$

  Using the closed form, FW = $1000(1.1^{30}-1)/0.1 = 164{,}494$

  Using EXCEL, it is FV(0.1,30,1000,0,0) = −164,494.

- Uniform Series Sinking Fund $A = F(A/F, i\%, n) = Fi/((1 + i)^n - 1)$

  Here, we do the inverse calculation; what would we need to set aside annually to achieve a fixed value in n years? In Business, this is referred to as a sinking fund. Suppose our engineer needs to have a million dollars for his retirement fund. Then we write:

Annual Contribution = 1000000(A/F, 10%, 30) = 1000000(0.00608) = 6,080.

Note the relationship of this to the uniform series FW; they are the inverse of each other.

Similarly, AC = 1000000(0.1)/(1.1^{30}–1) = 6,079.2

Or, using the EXCEL function PMT (with the same parameters as FV),

AC = PMT(0.1,30,0,1000000,0) = 6,079.2

- Uniform Series Present Worth PW = $A(P/A, i\%, n) = A((1 + i)^n - 1)/(i(1 + i)^n)$

We round out the discussion by looking at the Present Worth of a series of annual payouts. Suppose that an engineer needs $50,000 a year for the next 10 years of her retirement.

How much must she pay for such an annuity? We write:

PW = 50000(P/A, 10%, 10) = 50000(6.145) = 307,250.

Using our calculator, PW = 50000(1.1^{10}–1)/(0.1(1.1^{10})) = 307,228

Using EXCEL, we write PW = PV(0.1,10,50000,0,0) = 307,228

- Uniform Series Capital Recovery $P(A/P, i\%, n) = P(i(1 + i)^n)/ ((1 + i)^n - 1)$

Finally, we look at the reverse of the USPW. If we have to outlay a certain sum for a piece of equipment, what would we have to make per year to pay off that investment in n years? Suppose a personal loan from your father of $30,000 for down payment of a house and you want to pay it off in 10 annual equal payments.

Then we calculate Annual Payments = 30000(A/P, 10%, 10) = 30000(0.1627) = 4,881

Using the closed form, Annual Payments = 30000(0.1(1.1^{10}))/(1.1^{10}–1) = 4,882.36

Using EXCEL, Annual Payments = PMT(0.1,10,30000,0,0) = 4,882.36.

These are the fundamental tools that we will use in the following sections. Note that they are all "equivalent" in that they express the same relationships between the cash flow only from different perspectives.

4.4. Choosing Amongst Projects Using Equal Worth

How does an engineering organization select what projects will be done over the current year? Obviously, one would select the most financially attractive project. But how do we quantify that? Moreover, there are often other criteria that can come into play. Projects may be mandated by law and thus have to be done, regardless of the financial details. Sometimes too, projects are deemed strategic and are executed to gain a niche in the marketplace or to preempt a competitor from doing the same. Sometimes, the CEO just says "let it be done". For the purpose of Engineering Economics, we are going to assume that none of these other important categories of criteria exist. All our projects are considered equal in the sense that they are all equally valuable. Each can be staffed with the appropriate personnel, and each has the same risk associated with its execution. We shall assume the same MARR and inflation rate for all possible projects. Obviously, we have constructed a very simplistic model.

Parenthetically, we note the importance of the MARR. To illustrate this, consider the problem to select between a purchase and a lease, a very common engineering decision. The details are:

The local Fire Station wants to acquire a new fire truck. The authority can either purchase it for $400,000 or lease it at $25,000 per year for 24 years. The salvage value of the truck in 24 years will be $60,000. What should it do?

Suppose the MARR is 0. Then the lease would cost $24 \times 25000 = \$600,000$. If the Fire Station chose to purchase the vehicle, the cost would be $400,000 - 60,000 = \$340,000$. Thus purchasing would be preferable. Now let us assume a MARR of $i\%$. The defining equation becomes:

$$400,000 - 60,000(P/F, i\%, 24) \approx ?\ 25,000 + 25,000(P/F, i\%, 23)$$

(note that leases are paid at the beginning of the time period).

Let us try some values of *i* in Table 4.2:

Table 4.2. Experiment with MARR.

i%	400,000 − 60,000(*P*/*F*, *i*%, 24)	25,000 + 25,000(*P*/*A*, *i*%, 23)	Purchase-Lease
0	340,000	600,000	−260,000
1	352,744	534,500	−181,756
2	362,698	482,280	−119,582
3	373,720	415,500	−41,780
4	376,594	396,425	−19,831
5	381,394	378,700	+3,190

We can see that at about 4.85%, the two alternatives are equal. What this illustrates is that the choice of the MARR greatly affects the solution and hence the choice between the alternatives. The higher the rate, the less the present value of future cash flows.

Now we consider the problem of deciding between two projects, P_a and P_b. The most obvious way is to project all of the cash flows back to the present and compare their Present Worth values. Pick the greater as long as it is positive. In the literature, this is often referred to as the Net Present Value (NPV) approach. In our case, we use either the single *P*/*F* formula for each cash flow or the *P*/*A* one if regular payments are made. But we still have a problem in comparison. Are the lives of the projects equal? If they are, a simple comparison is valid. If not, we have to adjust the life span in some fashion. We examine three scenarios:

(1) Equal life spans
(2) One life span is a multiple of the other
(3) Unequal life spans

Scenario 1 — Equal life spans

In this case, the length of life spans is the same. We make a simple comparison.

Example for Scenario 1

Two projects are under consideration. Project A costs $500,000 to do and has a projected income stream for 3 years. Projected revenues are $150,000 in year 1, $200,000 in year 2 and $300,000 in year 3, with no salvage value at the end. Project B costs $450,000 and has a life of 3 years. Its income stream is $200,000 for each year, with a salvage value of $25,000. The company demands a MARR of 10%. Which is to be chosen, all other things being equal?

Solution:

$$\begin{aligned} PW_A &= -500{,}000 + 150{,}000(P/F,\ 10\%,\ 1) + 200{,}000(P/F,\ 10\%,\ 2) + \\ &\quad 300{,}000\ (P/F,\ 10\%,\ 3) \\ &= -500{,}000 + 150{,}000(0.9091) + 200{,}000(0.8264) + 300{,}000(0.7513) \\ &= -500{,}000 + 136{,}365 + 165{,}280 + 225{,}390 = 27{,}035 \\ PW_B &= -450{,}000 + 200{,}000(P/A,\ 10\%,\ 3) + 25{,}000(P/F,\ 10\%,\ 3) \\ &= -450{,}000 + 200{,}000(2.487) + 25{,}000(0.7513) \\ &= -450{,}000 + 494{,}924 + 18{,}783 = 63{,}707 \end{aligned}$$

Choose Project B, assuming that all other variables are equal. Note that we could also compute the Rate of Return (ROR) of the projects. A's is 27,035/500,000 = 5.4% and B's is 63,707/450,000 = 14.2%.

Scenario 2 — One life span is a multiple of the other

In this case, we assume that we can "repeat" the project with the lower life span. Economists call this the "repeatability" assumption; that all cash flows will reoccur in exactly the same way, for each repeated cycle.

Example for Scenario 2

Project A has a life span of 6 years. It will cost $400,000 and will generate $100,000 a year for those 6 years. At the end, it will have a salvage value of $50,000. Project B has a life span of 18 years and costs $400,000 to build. It will generate an income flow of $50,000 a year for those 18 years and will have a salvage value of \$100,000. We cannot compare the two projects directly if their life spans are not identical. Hence, we repeat Project A two times with exactly the same cash flow. Which should be chosen?

For the first life cycle, we draw the time diagram (Figure 4.2(a)).

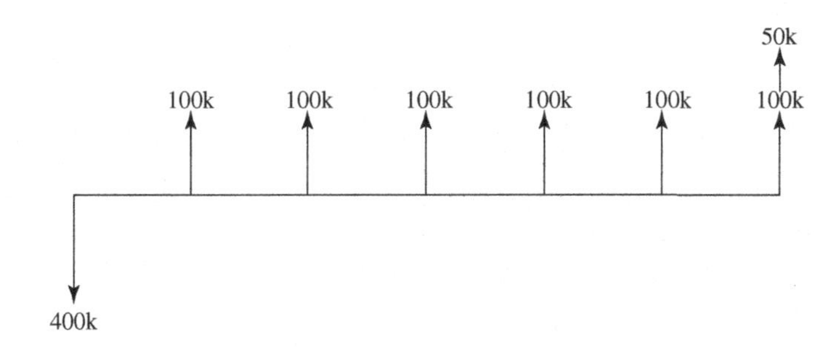

Figure 4.2(a). Time-line for P_A.

$$PW_{A1} = -400{,}000 + 100{,}000(P/A,\ 10\%,\ 6) + 50{,}000(P/F,\ 10\%,\ 6) = -400{,}000 + 435{,}500 + 28{,}225 = 63{,}725$$

Or to simplify, with the notation that A2 means the second running of exactly the same project (Figure 4.2(b)):

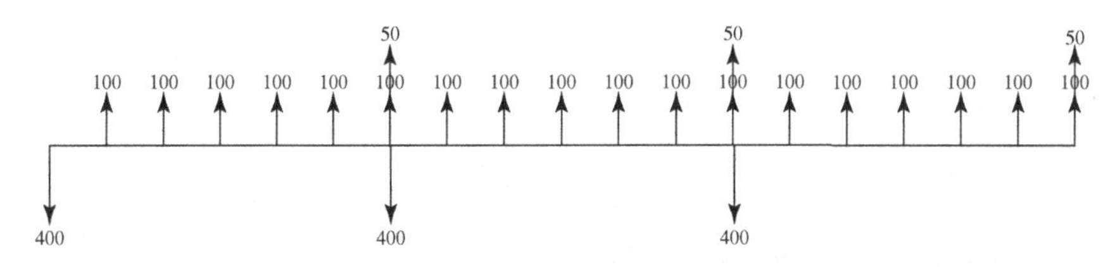

Figure 4.2(b). Replicated time-lines for P_A.

Thus we need to calculate the PW of A2 and A3. But the above formula only calculates the PW to the year 0 of the cycle (i.e., year 6 for A2 and year 12 for A3). We need to "bring their year 0" back to the real year 0.

Therefore,

$$PW_A = PW_{A1} + PW_{A2}(P/F,\ 10\%,\ 6) + PW_{A3}(P/F,\ 10\%,\ 12) = 63{,}725 + 63{,}725(0.5645) + 63{,}725(0.3186) = \$120{,}000$$

The PW of the second project is:

$$PW_B = -400{,}000 + 50{,}000(P/A,\ 10\%,\ 18) + 100{,}000(P/F,\ 10\%,\ 18) = -400{,}000 + 50{,}000(8.201) + 100{,}000(0.1799) = \$27{,}950$$

Thus choose Project A!

Scenario 3 — Unequal life cycles

Here we must resort to the Least Common Multiple or LCM method. We find the LCM of the two life spans and replicate each project for the requisite number of times.

Example for Scenario 3

Englebert Engineer has two offers to finance a new car. Car A has a down payment of $15,000, a yearly payment of $3,500 for 6 years when the salvage value is estimated to be $1,000. Car B has a down payment of $18,000, a yearly payment of $3,100 for 9 years and a salvage value of $2,000. Assume the usual MARR of 10%.

Solution:

The LCM of the spans of A (6) and B (9) is 18. Thus we need to repeat Car A's analysis three times and Car B's twice. Let us start with PW_A.

$$PW_{A1} = -15{,}000 - 3{,}500(P/A,\ 10\%\ 6) + 1{,}000(P/F,\ 10\%,\ 6) = -15{,}000 - 3{,}500(4.355) + 1{,}000(0.5645) = -\$29{,}678$$

As before, this must be repeated three times, and each value brought up to the present time.

$$PW_A = PW_{A1} + (P/F,\ 10\%,\ 6)PW_{A1} + (P/F,\ 10\%,\ 12)PW_{A1} = -29{,}678(1 + 0.5645 + 0.3186) = -\$55{,}887.$$

Similarly,

$$PW_{B1} = -18{,}000 - 3{,}100(P/A,\ 10\%,\ 9) + 2{,}000(P/F,\ 10\%,\ 9) = -18{,}000 - 3{,}100(5.759) + 2{,}000(0.4241) = -\$35{,}025$$

and

$PW_B = PW_{B1} + (P/F, 10\%, 9)PW_{B1} = -35{,}025(1 + 0.4241) = -\$49{,}879.$

Plan B is the better offer.

In conclusion, LCM is possible but can get ugly if the LCM is large (say 19 and 31 years!).

The second present worth approach is to calculate the Annual Worth of the project. The standard Present Worth calculation is done and then that value is "annualized" to an equivalent yearly value.

Let us redo the above example.

Example for Annual Worth of Two Projects

$PW_A = -15000 - 3500(P/A, 10\%\ 6) + 1000(P/F, 10\%,6) = -15000 - 3500(4.355) + 1000(0.5645) = -29{,}678.$

Now PW_A's Annual Worth is $-29{,}678(A/P, 10\%, 6) = -29{,}678(0.2296) = -\$6{,}814.$

Similarly, $PW_B = -\ 18000 - 3100(P/A, 10\%\ 9) + 2000(P/F, 10\%,9) = -\$35{,}025.$

And PW_B's Annual Worth is $-35{,}025(A/P, 10\%, 9) = -35{,}025(0.1736) = -\$6{,}080.$

The Annual Worth approach avoids the awkward LCM problem and expresses the value in an intuitive matter to all project members.

4.5. Other Methods

We now discuss other common methods of project selection. In most cases, we would use all of these methods for comparison.

4.5.1. Benefit-cost ratio

The benefit-cost ratio (BCR) is defined as:

$$BCR = benefit/cost$$

The higher the BCR, the more likely the project is to be chosen.

BCR > 1 project is profitable

BCR = 1 project breaks even

BCR < 1 project is going to lose money

Example for BCR

The City of Oshawa wants to improve the access from Stevenson Road to Highway 401. It has been the site of massive traffic slow-downs in the past. By spending $1.5 million dollars on a new over-pass, congestion would be significantly relieved. The City Engineer estimates that the annual benefits to travelers in saved time would be $225,000. The overpass has an expected life of 30 years, will require $65,000 annually for maintenance and will have a salvage value of $300,000. Using a MARR of 8%, calculate the BCR.

Note that we can solve this as a Present Worth analysis, as a Future Worth, or an Annual Worth analysis.

We use the Present Worth example.

$$PW_{\text{benefits}} = 225\text{K}(P/A, 8\%, 30) + 300\text{K}(P/F, 8\%, 30) = 2.563\text{M}$$

$$PW_{\text{costs}} = 1{,}500\text{K} + 65\text{K}(P/A, 8\%, 30) = 2.232\text{M}$$

$$B/C = 2.563/2.232 = 1.15$$

Therefore the project should proceed. How did the City Engineer come up with a user benefit estimate of $225,000 annually? We leave that as an exercise!

4.5.2. Internal rate of return

The Internal Rate of Return (or IRR) is the percentage interest rate where the net present value of cash inflow equates cash outflow. Consider the following PW example. A project costs $500,000 to do. It returns $100,000 in year 1, $200,000 in year 2, and $300,000 in year 3. What is its IRR?

Start by using trial and error. With an assumed MARR of 0, the PW is $100,000. If we use 10%, the cash flow is $481,592 and the PW = –$18,408. So, we are close. The 10% is too much, but not by much. Try 8%. PW = $2,210. Slightly too little. With 8.2%, PW = $87. Thus the IRR is about 8.2%. Using the IRR allows us to compare projects against their IRRs and is often used in business.

The higher the IRR, the better is the project to pursue. Nevertheless, one has to be cautious if IRR becomes unrealistically high. Remember the internet bubble that burst in early 2000. The IRRs for many startup opportunities were grossly inflated, not achievable and unrealistic, yet there were many naive believers who assumed that a high IRR meant a strong business case.

4.5.3. Payback

Payback is a simple yet common method of calculating the point of profitability. Measure the cost of the project against the benefits. At what point (measured normally in months) does the cost equal the benefits? This is the point of profitability. While useful in the executive office because of its simplicity, it ignores the time value of money. Also it ignores the future profitability that is likely to accrue. If Project A has a breakeven point of 3 months but Project B has one of 6 months, it is not necessarily true that Project A is better. What if all it delivers is one month of profit while B brings in huge amounts after 6 months?

4.6. Scoring Models

We conclude with an introduction to scoring models. We want to set up a model that can be used to rank several projects in a fair, unbiased manner. The idea here is to set up a model with which we can compare relevant criteria. Typical criteria might include the cost of the project, the profit earned by the project (often called the "payoff"), and the probability of success. But we could consider other things too, such as availability of people, equipment, perceived value to the organization and so on. The easiest model to construct is a simple linear model.

Let us use a simple example. You are taking a course as part of a five-course loading. Each course has a range of possible marks from 0 to 100. Your final term average is the sum of them divided by 5.

That is, if your mark in a course is m_i, then your $TermAverage = (m_1 + m_2 + m_3 + m_4 + m_5)/5$. That is a scoring model. Note that all of the courses have the same weight in this case. If you were applying to graduate school, the weights might well change. In general, if we had n variables, the score would be:

$$\Sigma m_i c_i \text{ for } i = 1 \text{ to } n.$$

where c_i's are the weighting factors, and we also require that the sum of the c_i equal 1 (we say that they are normalized), but you vary each c_i depending on its relative importance. When we form a sum, we include variables that we consider benefits. Clearly, profit and probability of success are benefits: the higher those values, the greater the score. But what about cost? It is a disbenefit: the smaller the cost, the greater the value to the scoring model. Thus we could construct a very simple scoring model connecting all three variables, namely probability of success (Psxs), payoff, and cost.

$$\text{SCORE} = \frac{(\text{Psxs} + 2 * \text{payoff})}{\text{Cost}}$$

We could even build a more complex model using other factors. Some might include:

- probability of technical success
- market size
- availability of staff
- strategic positioning
- favourability of regulatory environment
- probability of market success
- market share
- degree of organizational commitment
- organizational alignment
- skill-set of team members

For each of these variables, we need an algorithm for ranking each criterion in a fair and repeatable manner. For example, in setting a value for "skill-set of team members", we might use the following scoring guide for assigning a score from 1 to 10.

10. All skills ample with redundancy
9. All skills ample no redundancy
8. All technical skills available
7. Most technical skills available
6. Some technical retraining needed
5. Some professional retraining needed
4. Extensive technical retraining necessary
3. Extensive professional retraining
2. All technical personnel must be hired
1. All of the team must be hired

In setting scoring values, we often must assign scores to attributes that are not directly quantifiable. As a silly example, suppose that one of our criteria was the "attractiveness of the project manager". Such problems often come up in Social Science. A psychologist named Likert defined what he called the Likert Scale. You assign a score from 1 to 5, integers only. So a 3 would be a person who is average looking; neither handsome nor ugly. A 5 would be drop-dead gorgeous while 4 would be in between 3 and 5. Similarly, a 1 would be notably unattractive. A 2 would be not handsome but not really ugly. Likert's insight was that we can make those 5 gradations quite reliably. Using 10 is much more problematic. An example when you might use such a scale is in evaluating the user's perception of the goodness of a piece of software. You might ask a question like "how user-friendly is the Vista Operating System?" Ask a thousand users and average their score. Then repeat on a previous system like Windows-XP and you would have a good measure of improvement or not. Likert was following Galileo who famously said "what is not measureable, make measureable". Please put this into your Project Management kit bag.

Finally, we conclude by noting that scoring models are common. They are relatively easy to use and we will see them again in our journey. They are also bias-free, transparent, and defensible. But they can give a false sense of security, implying a precision that is not there. Sensitivity analysis should be conducted to determine how the variables affect the project outcome.

4.7. Chapter Summary

We have discussed several techniques for project selection. Doing a solid, complete economic analysis is vital for project selection and success.

4.8. Discussion Issues

1. You are in charge of justifying the extension to the 407 highway, a major super highway in Ontario. The project will cost hundreds of millions. But the expected benefit is that the average transit time will be reduced by 30 minutes. You need to quantify the benefits of the upgrade. How would you do this? You must come up with an annual benefit dollar value. State all of your assumptions and conclude with a dollar value.
2. Same upgrade. There are also disbenefits associated with the above proposal. Many farmers will have their farmland expropriated. Quantify the initial cost of those disbenefits. State all assumptions.

4.9. Multiple Choice Questions

1. An IRR of 0% means that
   (a) This project must be selected
   (b) The expected benefits of the project do not equal the cost of the project
   (c) The MARR is also 0
   (d) If this project were selected, it would not be on economic grounds

2. This is NOT a factor in setting the MARR:
   (a) Overnight Lending Rate
   (b) Risk of the project
   (c) Capability of the company to do the work
   (d) Capitalization of the company

3. Sensitivity analysis describes:
   (a) Organizations' ability to raise liquid cash
   (b) The credit rating of an organization
   (c) The risk aversion of the customer
   (d) The identification of the input variables that will most affect the outcome of the project

4. Closest future worth in year 10 of $10,000 at an interest rate of 10% is:
   (a) $10000(1 + .1)10$
   (b) $10000(F/P, 10\%, 10)$
   (c) $10000(P/F, 10\%, 10)$
   (d) $(F/P, 10\%, 10)$

5. Public Sector projects do not have "profit". Instead they replace increasing profit with:
   (a) Maintaining high quality
   (b) Minimizing costs
   (c) Sole-sourcing the costs
   (d) Paying taxes

6. In our cash flow problems, which of the following is NOT an assumption?
   (a) Interest rates are constant
   (b) Positive cash flows are made at the end of time periods
   (c) Negative cash flows happen at the beginning of time periods
   (d) Positive and negative cash flows can happen at any point

7. If you had a 20-year annual annuity flow that begins 6 years from now, then you would use which functions to compute its total PV?
   (a) $(P/A, i\%, 26)$
   (b) $(P/F, i\%, 6)(P/A, i\%, 20)$
   (c) $((P/A, i\%, 20)\ P/F, i\%, 6)$
   (d) $(P/A, i\%, 20)$

8. The department wants to replace its PACE computers in 4 years. The cost then is estimated to be \$100,000. Which expression states the amount for this that must be set aside in this year's budget, assuming a 3% economy?
   (a) $100{,}000(P/F, 3, 4)$
   (b) $100{,}000(F/P, 3, 4)$
   (c) $100{,}000(F/P, 4, 3)$
   (d) $100{,}000(P/F, 4, 3)$

9. If you wanted a 3-year annual \$1,000 annuity flow that begins 6 years from now, then you would use which functions to calculate the amount you would have to spend today to get that flow? Assume 5%.
   (a) $3(1000(P/F, 5, 6))$
   (b) $1000(P/F, 5, 6) + 1000(P/F, 5, 7) + 1000(P/F, 5, 8)$
   (c) $3(1000(F/P, 5, 6))$
   (d) $1000(P/F, 5, 5) + 1000(P/F, 5, 6) + 1000(P/F, 5, 7)$

10. My tuition at university was \$300 in 1961. Your university tuition is \$7,000 this year. In real terms who paid more, you or me?
   (a) We paid about the same
   (b) I paid more
   (c) You paid more
   (d) You paid a LOT more

11. In 1626, the Dutch bought Manhattan Island for $24. Assuming an average inflation rate of 10% what would it be worth today (ballpark)?
    (a) A whole lot
    (b) A billion dollars
    (c) A trillion dollars
    (d) A quintillion (10^{15}) dollars

12. A project's payback period ends when:
    (a) Maximum profit is realized
    (b) Unit profit is realized
    (c) Monthly revenue exceeds monthly cost
    (d) Cumulative revenue equals cumulative cost

13. A company wants to replace a machine in 15 years that will cost $10,000. It does this by setting up a sinking fund of $300 per year. What is the closest interest rate that will make this happen?
    (a) 10%
    (b) 10.5%
    (c) 11.2%
    (d) 11.6%

14. A product is estimated to generate $1 million next year and $2 million the year after. What is the net present value of the total amount to be generated by the product over the next two years, assuming 10% interest rate?
    (a) $1 million
    (b) $2.56 million
    (c) $2.81 million
    (d) $3 million

15. If the MARR were zero, we might conclude:
    (a) Nothing
    (b) The time-value of money is meaningless

(c) We are doing the project for non-economic reasons
(d) We are doing the project for economic reasons

4.10. Exercises

1. You have just won the "Million Dollars for Life" lottery which pays you $1,000,000 annually for 25 years. But the Ontario Lottery and Gaming (OLG) wants to offer you a lump sum of $20M. Assuming a MARR of 10%, Deal or No Deal? Show all work.
2. You have just been made Dalton McGuinty's Economan (an ex-premier of the Province of Ontario). Your first task is to justify the financial cost of Ontario's Family Day. Without factoring in the soft costs (such as: improved productivity because of happier Ontarioans), how would you get an in-the-ballpark estimate of the cost of the proposal? Assume that all working people have the day off. KISS folks. (Keep It Simple Stupid principle).
3. Using the 10% table provided in Appendix C, what is $10,000 worth 5.2 years from now, at an assumed interest rate of 10%? Show your work.
4. Ellie and Elliott Engineer each deposit $1,000 every year into their retirement savings account, each bearing 10% interest.
    - Draw the Cash Flow diagram.
    - How much will Elliott have in 30 years?
    - If Ellie wants to have $1,000,000 then, what would her interest rate have to be?
5. A project has the following costs and benefits. What is the payback period?

Year	Cost	Benefit
1	1,400	0
2	500	0
3	300	400
4 to 10	0	300/year

6. Project A costs $200K. In year 1, it will earn $100K, $400K in year 2, $100K in year 3, and then never earn a cent. Project B costs $400K. It earns $100K in year 1, $200K, $300K, $200K, and finally $100K in subsequent years before becoming worthless after year 5.
   - Which project would you recommend and why? Assume a MARR of 10%.
   - At what interest rate would the two be identical?
7. How would you estimate the benefits to Oshawa drivers of the improvement to the Stevenson Overpass project?
8. Your Pointy Hair Boss (PHB) refuses to accept your Charter yet will not let you resign from the project. What is your next step?
9. How would you as a PM with a 5-year project, estimate the Overnight Lending Rate for the next five years and why would you care?
10. What does an IRR of 18% mean?
11. PW = $5000(P/F, 15\%, 5)(P/A, 10\%, 4)$ is the solution to what problem?
12. Sam and Cam are two hostile brothers. Their father, Bam, decided to use real estate in an attempt to get them back together. He gave them a house in Toronto 30 years ago. The initial price was $100,000. Cam, being a useless piece of driftwood, ignored the house altogether. Sam rented the house out at $1000 per month and split the money with Cam. Twenty years ago, he put on a new roof ($10,000). Ten years ago, he bricked the place ($50,000). Today, Cam has died and the place is sold for $1,000,000. What portion should Cam's widow get, assuming a 10% economy?
13. Sydney Crosby and his Penguins win the Stanley Cup. He renegotiates his contract. The Penguins give him a choice; take either a flat sum of $25 million dollars on Jan 1, 2012 or $10 million per year for the next 3 years, each amount payable at the end of each year, starting with 2011. Assuming a 10% MARR, which should he choose and why?
14. Project A promises income of $1,000 in year 2, $2,000 in year 3 and $4,000 in year 4; Project B promises income of $1000, $2000 and $4000 in years 0, 1 and 2 respectively. The total income for

both projects is $7000. Which project is more beneficial based on PV calculation?

PV exercise (i = 10%)	Project A		Project B	
	FV	PV	FV	PV
Year 0			1000	
Year 1			2000	
Year 2	1000		4000	
Year 3	2000			
Year 4	4000			
net present value				

15. Project A requires investment of $3,000 this year, $2,000 in year 1, and $1,000 in year 2; and it promises income of $1,000 in year 2, $2,000 in year 3, and $4,000 in year 4; what is its IRR? Similarly, Project B requires investment of $3,000, $2,000, $1,000; and promises income of $1,000, $2,000, and $4,000 in this year, year 1 and year 2, respectively; what is its IRR? Both projects have total income of $7,000 and investment of $6,000. Which project is a better one to pursue based on IRR calculation?

IRR exercise	Project A		Project B	
	Cash inflow	Cash outflow	Cash inflow	Cash outflow
Year 0		3000	1000	3000
Year 1		2000	2000	2000
Year 2	1000	1000	4000	1000
Year 3	2000			
Year 4	4000			
IRR				

References

1. Franklin, B. Advice to a Young Tradesman. 1748.

CHAPTER FIVE

Requirements and Scope Management

5.1. The Relationship between Requirements and Scope

Scope is essential to Project Management but what IS it? It is the summation of all of the steps and deliverables that we will do and create as we run the project to completion. But how do we determine them? Let us see what we have so far. We have the Charter, which is both the authority for running the project and a preliminary description of the scope. We have identified our stakeholders. Now we need to refine the Charter to produce the Project Scope Statement (PSS), one of the key project artifacts. But before we do that, we need to understand the qualities of the product we are going to create. For example, the steps necessary to build a functional nuclear reactor would be far more numerous and complicated than those necessary to write a piece of software to convert integers to Roman Numerals. We need to clearly state the requirements of the product, so that we do the minimum number of steps necessary to create that product. Remember, the other two sides of the Iron Triangle or Triple Constraint, are cost and time.

Requirements are the characteristics that the product must have; scope describes the steps and attendant deliverables necessary to create that product. It follows that our first task is to specify what our requirements are. Knowing that, we can invoke the necessary steps to achieve that quality. Thus we build the Requirements Specification Document (RSD) and then construct the Scope

Statement. These two documents will then guide the writing of the Project Management Plan.

The Scope Statement and the RSD are thus two closely related artifacts. The Scope Statement is a description of the work that must be performed to deliver a product, service or result with the specified features and functions. The definition of the "specified features and functions" is supplied by the RSD.

5.2. Requirements Management

The major goal of Requirements Management is to produce the RSD. This is a critical document. As we shall see later, the definition of a "defect" will be "a violation of a requirement". But what if the requirement is wrong or worse, missing? Can you imagine the Prime Minister of Canada going to the opening ceremony of the Confederation Bridge, and before cracking the traditional bottle of champagne over the concrete, declaring "*Mon Dieu*, this will not do! I want 4 lanes, not these two!" That would be a massive requirements violation, necessitating taking down the existing structure and rebuilding it from scratch. Now in Civil Engineering, we would not make such an obvious blunder, but in other areas of Engineering, especially Software, such mistakes are all too common.

5.2.1. Introduction to the problem

What are requirements? Why are they important? And why are their definitions so difficult? We start with the formal PMBOK definition: a requirement is "a condition or capability that is required to be present in a product, service, or result to satisfy a contract or other formally imposed specification" [1].

As a trivial example, suppose our project is to write a little piece of software that will accept an integer as an input and convert that integer to a Roman numeral form or issue an error message. The package is to be called "CRN" and to be invoked as a C runtime package by coding "CRN(N,STRNG)", with the semantics that, if the translation is correct, CRN will return a value of "0" and

STRNG will point to the Roman numeral equivalent of N. If there is an error, it will return the value of "−1", with STRNG pointing to the error message (Romans did not know about zero or negative numbers. They also had an upper limit of a million). Thus a requirement could be:

R1: input an integer between 1 and 1,000,000 in integer format

Now we see the difficulties. Is "1" valid or do we start at "2"? What does "integer format" mean? Developers could choose different interpretations. So a better statement might be:

R1: input an integer between 1 and 1,000,000 (inclusive) in integer format (two's complement 32-bit form).

Requirements are hard, for several reasons.

1. *Users do not know what they want*

If you are building a product that has never been developed before, how can the user imagine what the final product might look like? Think about specifying the requirements of a Windows-Apple icon environment when the only tool you have used was a punch card.

2. *Users and developers do not speak (or understand) the same language*

Years ago, software engineers used to separate out requirements from specifications. The idea was that non-technical stakeholders would specify the requirements and throw them over the wall into the developers' pen. The developers would then map the informal requirements into specifications. Imagine an insurance application. The requirement might be: "get the insured's age". The software engineer might map that into a two-character field, with legal values from 00 to 99. But the insurance agent might want "age" specified as the "number of days the insured has lived". Now good practice dictates that both the stakeholders and the developers construct the RSD together, each side explaining clearly any discrepancy in terminology. Indeed the IEEE standard for Software Requirements is called the Software Requirements Specifications (SRS) document, to reflect the marriage of the two approaches.

3. *There is no universal way to describe requirements*

How can we specify requirements? The obvious way is to use natural language. But natural language is inherently ambiguous. One can also use state diagrams, timing diagrams, Petri nets, mathematical notations, etc. Each will have its advantages and disadvantages, as we shall see.

4. *Natural language is inherently ambiguous*

Consider the tabloid's headline "Charles Drives Diana to Drugs". How to make sense of that snippet? Those of us who are "royal-followers" know that the reference is likely to be Prince Charles and Princess Diana of Great Britain's royal family. But maybe it is Charley Brown and Diana Krall that the snippet refers to. Or, perhaps, Charles berated Diana so badly that she got a splitting headache and had to go to the chemists for some Aspirin. Or maybe they took a nice drive in the Rolls to a little village called "Drugs". You get the idea.

5. *It is human nature to start doing the work without requirements, filling them in later*

Projects excite us and we want to get started as early as possible. Doing the engineering is what we love, not talking to the stakeholders! We can always change things later when the requirements are fixed.

Requirements are important. As we have said, the first point to understand is that they will be used for defect definition. They will also be used in quality control, both in the product and in the running of the project. Most software processes use a process called "Verification and Validation" (V&V). Consider the completion of a part of the Waterfall model used to write the code for CRN. The input to the coding process is the Detailed Design Document (DDD) and the output, the complete code module, unit-tested, as shown in Figure 5.1.

Verification means checking to see that we did what the inputs said we should do (doing the thing right). Validation means

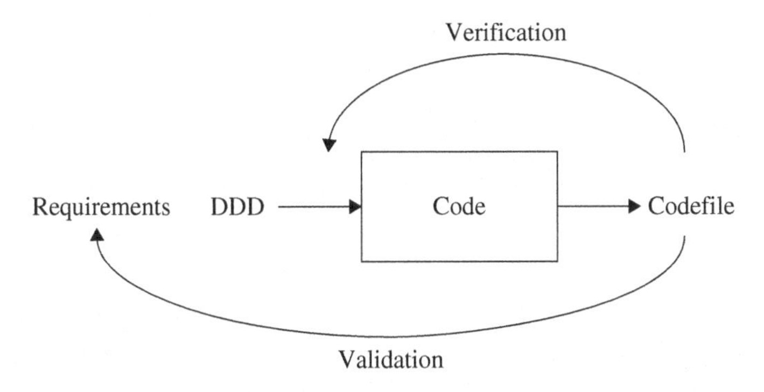

Figure 5.1. Verification and validation.

checking back against the Requirements to ensure that we are still meeting the requirements (doing the right thing). Verification is vital in controlling the execution of the project (see Earned Value in Chapter 7). But Validation is also very important, to be sure we are still meeting the prescribed conditions. This is so critical in Quality Control that organizations often use external evaluators to perform the V&V. This is often called Independent V&V (IV&V). In the Military, for example, IV&V is mandated to make sure the validators have no axe to grind.

This is an example of "is your baby beautiful" syndrome. In the fall, farming communities often have local fall fairs where farmers show off their produce. In addition, they often have friendly competitions such as the "most beautiful baby" contest. Now the judges of that contest cannot have their own children in the contest. Folks, there are a lot of ugly babies out there but not to their parents. For such contests, as with IV&V, it is important that there be no conflicts of interest in the judgements. Having the team members evaluate their own work is like having a parent choose the most beautiful baby from a group containing their own child. The developers will either ignore shoddy work (as it would be very embarrassing to have to reject your own work as unacceptable) or they will not see it (for if they could have seen it, they would have fixed it).

Finally, Requirements are important because defects in requirements can be expensive, very expensive. Several studies have shown that it is the most expensive phase in which to create and remove a defect. It has been estimated that a requirements defect that escapes detection at the requirements phase and is not fixed until Maintenance can cost 200 times the initial phase detection and correction. A USAF study [2] found that 41% of their defects were found in Requirements. Tom DeMarco reported 56% of all defects in one of his case studies were requirements related. Big companies like Raytheon and Boeing found that their reworks costs charged to requirements were 40% and 85%, respectively [3]. The famous Chaos Report [4] attributed 40% of detected defects to Requirements defects. This is logical in that if you have a Requirements defect, and it escapes phase-containment, the eradication costs escalate rapidly for each phase it goes through undetected; of course, since you are designing and coding to incorrect specifications.

5.2.2. Requirements life cycle

The Requirements Life Cycle begins with the acceptance of the Charter and the establishment of the Stakeholders (listed in the Stakeholders Registry). These are the inputs to the Gathering Requirements phase. The outputs of this process are the Requirements Specification Document (RSD), the Requirements Management Plan (RMP), and the publishing of the Requirements Traceability Matrix (RTM). These are then updated by the Requirements Change Process, a subset of the Monitoring and Controlling phases of the Project Life Cycle.

5.2.2.1. *Gathering requirements*

The Requirements Team (RTeam) is formed, likely headed by the Project Manager. The RTeam may use several methods to "elicit" the requirements (this is called the Elicitation Phase). Requirements can be categorized into similar baskets (see Section 5.2.3). Within those baskets, structuring can be done. Requirements come from the stakeholders; it is the primary duty of the RTeam to map stakeholder expectations into Requirements. Then the results will be put into a structured format to form the RSD. Typically, the RTeam will then build a prototype and have the stakeholders "play" with it. This experience will form a feedback corrective loop and will lead to corrections to the RSD. Finally, the RSD will be baselined and the process will flow into the next phases.

There are several ways of gathering requirements including, but not limited to, the following:

1. Interviews
2. Focus groups
3. Facilitated workshops
4. Group facilitation techniques
5. Group decision-making techniques
6. Surveys
7. Observations
8. Prototypes

Now we quickly explore each technique.

1. Interviews

This involves the RTeam interviewing all stakeholders, previous project participants, and subject-matter experts (SMEs). The team may use predefined questions or expand those with spontaneous ones. All responses should be recorded.

2. Focus groups

These are more structured interviews, where a leader will guide stakeholders and/or SMEs through the questions. The emphasis is to encourage synergy amongst the participants rather than a one-to-one approach.

3. Facilitated workshops

These are focused sessions bringing together key stakeholders to thrash out requirements. Typically, stakeholders have orthogonal goals: the Chief Financial Officer (CFO) wants costs contained, the

Quality Boss wants superb quality, the Sales Manager wants a glitzy product, and so on. Here they can hear and respond to conflicting objectives in real-time and appreciate alternate concerns. In software, these were developed by IBM and the Canadian Government (called Joint Application Development sessions (JADs)) and have a very high track record in reconciling differences between users and developers. In manufacturing engineering, these are called examples of Quality Functional Deployment (QFD) techniques.

4. Group facilitation techniques

Group activities can be structured in precise ways to focus the group on exploring various regions of requirements elicitation. Examples are:

- Brainstorming

People sit in a circle and toss around wild ideas to explore unknown regions of thought (outside of the box thinking).

- Nominal group technique

Brainstorming followed up by a group vote to prioritize the various ideas.

- Delphi technique

The general problem is presented to several SMEs. Each one then goes away to a small room and works out his set of requirements. They then all reconvene and present their personal findings with comments. They then disperse and start another cycle. Typically, four iterations are enough for consensus.

5. Group decision-making techniques

This is a way of prioritizing various requirements, and can be applied to any of the above techniques.

- **Unanimity** — all agree on the prioritization
- **Majority** — more than 50% of the group agree
- **Plurality** — the largest block prevails even if not a majority
- **Dictatorship** — one individual makes the decision for the group

6. Surveys

Surveys and questionnaires can be used to obtain fast responses from a large group.

7. Observations

Here a member of the team "shadows" a worker to observe hidden requirements. It is a useful technique when individuals cannot articulate their requirements.

8. Prototypes

A fake-up of the final is presented to users for their evaluation. Note that in manufacturing, a prototype is a one-off real artifact before mass production of the artifact. The prototype of a Boeing 777 would be the first flyable version of the aircraft. In Software Engineering, a prototype is really a mockup of the original. A mockup of the 777 might be a paper-maché model which would show the size and shape of the final product but it certainly could not "fly".

5.2.2.2. *Requirements specification document*

This is the document that describes the product's characteristics. It is likely the most important document of the entire project. It may take one of three general forms: informal, formal, or mathematical. The informal Requirements Specification Document (RSD) is a brief outline of the requirements using a few simple paragraphs or diagrams. The informal form would be used for student exercises, scaffolding projects, programs that had little risk consequence if they failed. The formal document follows a prescribed format that is well-documented. Using the IEEE SRS (see below) is a typical example. An organization might well develop its own procedure for requirements. Finally, the mathematical approach uses either a formal notation or a language suited to the problem at hand. For example, the telecommunications industry has developed languages for expressing and formally checking computer communications protocols (languages such as LOTOS, SDL, Estelle, and MONDEL). NASA has also used formal methods for describing their space shuttle software.

Whatever the form, the RSD must be easily traceable. That is, each requirement must have a number. Normally, requirements are hierarchical and the numbering should reflect that. The stated requirement should point back to the stakeholder that requested it. Finally the RSD must be signed-off. That is, a competent group of examiners must agree (and sign that agreement) that the requirements are complete, well-organized, clear, and unambiguous. Each requirement should have documentation covering at least the following attributes:

- have benefits that outweigh the costs of development,
- be important for the solution of the current problem,
- be expressed using a clear and consistent notation,
- be unambiguous,
- be logically consistent,
- lead to a system of appropriate quality,
- be realistic with available resources,
- be verifiable,
- be uniquely identifiable, and
- does not over-constrain the design of the system.

How much detail should one provide? That depends on the size and scope of the project. Some general considerations include:

- the size of the system,
- the need to interface to other systems,
- the readership,
- the stage in requirements gathering,
- the level of experience with the domain and the technology, and
- the cost that would be incurred if the requirements were faulty.

5.2.2.3. *Requirements management plan*

The Requirements Management Plan (RMP) is the physical document that describes how the requirements are to be gathered, analyzed, described, verified, and controlled throughout the span of the project. Based on stakeholder needs, wants, and expectations, requirements can be collected and generated. The phase-to-phase relationship (sequential, overlapped, or rolling-wave) will dictate the time-line for requirements management.

The chosen relationship will be documented in the RMP. Components of the RMP contain at least the following:

- methods of requirements eliciting,
- time-lines for elicitation,
- monitoring activities (IV&V),
- configuration management of requirements (see below),
- requirements prioritization process (see below),
- product metrics that will be collected and why they were chosen, and
- traceability structure.

5.2.2.3.1. Establishing the requirements change process

Once the RSD has been signed-off, it is placed under Configuration Control. We could just point you to Section 5.2.6 and have you substitute, *mutatis mutandis*, the appropriate requirements language, but since it is so important to get the requirements right, we detail the Requirements Change Process steps here.

1. Someone sees a necessary change. She fills out a Change Request (CR) form, detailing the change. Requirements can change because of:
   - the business process changes,
   - technology changes, or
   - the problem becomes better understood.

2. The CR is sent to a person capable of assessing the time and costs of the change (likely the Project Manager (PM)). He first qualifies the request to ensure that it is worthy of evaluation. If so, he appends his time and cost estimates.
3. The Change Request is sent to the Change Control Board (CCB) or Change Review Board (CRB). A meeting of the CRB is then scheduled. The CRB consists of (at least):
   - the Project Manager
   - a technical person
   - a cost expert
   - a marketing person
   - the customer
   - a quality person
4. The CRB meets and approves or rejects the request. If the former, the changed PMP is baselined and all copies are updated. If it is refused, the reasons are recorded and sent to the initial requester.
5. All details of the CR are recorded in the Requirements Database.

Requirements analysis is a continual process. The benefits of changes must outweigh the costs. Certain small changes (e.g., look and feel of the User Interface, UI) are usually quick and easy to make at relatively little cost. Larger-scale changes have to be carefully assessed. Forcing unexpected changes into a partially built system will probably result in a poor design and late delivery. Some changes are enhancements in disguise. Avoid making the system *bigger;* rather only make it *better.*

5.2.2.3.2. Establishing the relative priority of requirements

Stakeholders will have different rankings for priorities. The Chief Financial Officer (CFO) rates cost as the top priority while the Quality Chief rates reliability as more important than cost. The Project Manager as part of her stakeholder management, must list all of the priorities in sequence with the stakeholders across the top. The stakeholders then rank the importance of the priorities. Of course, they will conflict. The PM then has to set the priorities as she sees them, and revisit each stakeholder to explain the divergence of the priorities chosen and why. This reconciliation is vital for stakeholder cohesion.

5.2.2.4. *Requirements traceability matrix*

The Requirements Traceability Matrix (RTM) is a table with the requirements listed as rows and important evaluation points across the top as columns. The most important entry is the stakeholder who defined the requirement. Then as IV&V progresses, each "pass" would be noted as that requirement is rechecked for acceptance. Thus we can track the progress of requirements through the execution of the project. The tracing, in addition to the actual meeting to the requirements, can also be linked to:

- business needs (sales)
- project objectives
- project scope
- product design and development
- test cases

5.2.3. Types of requirements

Requirements fall into two broad categories: non-functional and functional. Non-functional requirements express constraints that must be adhered to during the project (e.g., all deliverables must be inspected according to procedure 123). Functional requirements express actions that the system must perform (e.g., the system must shut down within 1 second when the Control-Z key is depressed).

5.2.3.1. *Non-functional requirements*

There are three classes of non-functional requirements (NFRs):

1. Categories reflecting: usability, efficiency, reliability, maintainability, and reusability (requirements that constrain the design

to meet specified levels of quality). Examples include at least the following:

- response time
- throughput
- resource usage
- reliability
- availability
- recovery from failure
- allowances for maintainability and enhancement
- allowances for reusability

2. Categories constraining the *environment and technology* of the system:
   - platform, hardware, software, communications
   - technology to be used
3. Categories constraining the *project plan and development methods*.
   These are commonly placed either in the contract or the PMP. Examples are:
   - development process (methodology) to be used
   - explicit charter of accounts
   - cost and delivery dates

5.2.3.2. *Functional requirements*

Functional Requirements (FRs) have metrics associated with them. While it is hard to specify if a NFR has been met (e.g., how do you certify "must be easily maintained"?), the specification of functionals is clear. The condition is met or it is not. If the functional requirement is <print "hello world"> within 1 second, we can measure that and confidently assert whether it has been met or not. FRs specify what the form of inputs is. They describe the syntax and semantics of the outputs to be created. If data is to be collected from other applications, that data is described by FRs. The nature and accuracy of computations to be performed are described. The timing and synchronization of computed data are specified by FRs.

In short, FRs are the good kind of requirements; NFRs are the bad kind. Whenever we can, we try to map NFRs into FRs.

5.2.3.3. *Requirement classifications*

Another approach is to subdivide requirements into a disjoint set of categories or baskets. This categorization is arbitrary; we chose a particular subdivision that makes sense to us; the IEEE SRS uses a slightly expanded group. Note that the requirements can be either NFRs or FRs.

REQUIREMENTS SUBDIVISION

1. Physical Environment
2. Interfaces
3. User and Human Factors
4. Functionality
5. Documentation
6. Data
7. Resources
8. Security
9. Quality Assurance

Let us look at each in sequence.

1. Physical Environment

Suppose that we are describing a laptop. What physical limitations should we be concerned about? For example, most laptops are rated to perform between 0°C and 40°C. So it is not a good idea to take it along on your next climb up Everest. More prosaically, do not leave it in the trunk of your car in a good Canadian winter, or on the shelf under its back window in the summer. There are other concerns too: magnetic interference, weight, electrical, humidity, water-resistance, drop height, and so on.

In addition, where is the product to be delivered? How many locations, and what quantities at each?

2. Interfaces

Is there input entering the system from an external source? What is its format physically, electrically and logically? Does the system generate data to be used by down-stream systems? Does the data need to be formatted? What is the medium of interchange?

3. User and Human Factors

What is the nature of the people (or machines) using this system? For example, in software usage, we usually identify three classes (or modalities) of users: casual, advanced, and maintainers of the software. Imagine the differences between the user manuals for each class. Will there be several types of systems? Blackberry and Apple tablets come in three models, for example. What skill level is assumed in using the system, in either modality or option sizes? What kind of training is needed for each class of users? How easy is it to use the system for "normal" users? Can challenged users use it (people that are near-sighted, colour-blind, blind, handicapped, illiterate, mentally-challenged)? How difficult is it to misuse the system? Can the system be restarted easily?

4. Functionality

These are the normal FRs that we have discussed. What will the system do? When will it do it? How long will it take? Are there constraints on processor speed, response time, or throughput? Are there different modes of operation? How and when can the system be changed (upgraded, modified, down-graded)?

5. Documentation

How much documentation is needed? As we shall show for Software Engineering, we can estimate this. Do we need to use the three modalities approach here too? How do we "certify" the quality of the documentation? For example, we could use the Gunning Fog index: the level of English complexity, in terms of years of schooling = (average number of words/sentence + percentage of words with 3 syllables or more) × 0.4. *The New York Times* is 11–12 years. As popular wisdom has it, we should be aiming at a 5 level! We could also run experiments of students being given the documentation and a task to run and measure the completion time. That would be taken as a measure of the goodness or badness of the documentation. We note in passing that many high-tech products do not come with written documentation at all, only pictures!

6. Data

What should the format of the data be? How accurate must it be? How often will it be sent/received? To what degree of precision must calculations be made? Can we tell when the data becomes meaningless or as we used to say in FORTRAN, can we tell when we are wallowing around in round-off errors? How much data must flow through the system and are there temporal patterns that we must handle? A common cause of Web failure is an unexpected urge in incoming requests for server interaction. On the day before 9/11, the American Red Cross donation server collected $3,000; on 9/11 it collected over $1,000,000 before it crashed! Somebody missed an important requirement there.

7. Resources

What materials, personnel, monies are required to build, operate, and maintain the system? What skill-sets are needed of the developers? How much physical space will the system occupy? What are the requirements for power, air conditioning, heating, etc.? Is there a fixed timeline for project completion? Think about Y2K for software millennial projects. Is there a cost ceiling?

8. Security

Must access to the system data be controlled? Is the users' data protected as per legislation? Is the users' data protected from unauthorized access or accidental erasure and/or modification? Are the users' programs isolated from each other and the main application? Is the system protected against hacking attacks over the Internet? How often should the data be backed up? Are the backups protected against unauthorized access or physical disasters such as fires, earthquakes, public insurrections? All of these potential risks need to be expressed in requirements.

9. Quality Assurance

The following attributes need to be expressed in quality requirements:

- reliability
- availability
- maintainability
- security
- user evaluation
- other quality issues

Must the system detect and isolate faults? What is the acceptable Mean-Time-To-Failure (MTTF) and Mean-Time-To-Recover (MTTR)? How easily can the system be ported to new environments? Many of these issues are NFRs of course.

5.2.4. Characteristics of good requirements

What are the characteristics of good requirements? Here is a (non-exhaustive) list. Each requirement must be:

1. Unambiguous — requirements must be clearly expressed with no possibility of misunderstanding. Example: "the user must input a number". The user types in "three". Or a close approximation to the total number of atoms in the Universe! A better statement is "the user must input a number, specified by integers, greater than 0 and less than 1000".
2. Measureable — wherever possible, we must be able to quantify the requirement with a number. "R1: the phone must weight less than 0.5 kg, measured at sea-level". Even NFRs can be measured using a Likert scale as defined in Chapter 4.
3. Free from "motherhood" expressions

"Motherhood" expressions are pious or flatulent statements such as "the software must have high quality". As we shall soon see, quality depends on the application. Engineers sometimes call these "artsy" statements. One of our favorite baby books is called *Pat the Bunny*. One of the pages has the quote "how big is bunny" and the response is "bunny is SOOOO big". That is a motherhood response. The engineer of course would respond "bunny is 23.5 centimeters long and weighs 2.3 kilos".

4. FRs and NFRs remain unmixed

Functional requirements must be kept separate from non-functionals.

5. Free from design directives

Do not tell the implementer how to do her job. Let us illustrate this with a horrible example. A PM colleague told one of us this story. He was implementing a client-server network for a Government client who had just discovered "requirements". So the Client wrote a very extensive requirements document which included, amongst other things, the requirement that the client stations should be powered by Intel 386 chips, running at 33 MHz. When it came time to build the client stations, our colleague approached Intel and asked for a couple hundred 386 parts. Intel responded that they no longer sold that part but the 486, running at 50MHz and executing at 40MIPS (million of instructions per second), was much better than the 386 (11.4MIPS) and was the same price. Our colleague said no, he needed the 386 part, because that was what the requirement was! So at a much greater cost (Intel had to retool the line to make the obsolete part), our colleague satisfied the requirements. In reality, the Client should have specified the functionality in MIPS and left the part selection up to the contractor. Does the Client really care if a 486 or 30,000 Irish leprechauns are inside the stations, as long as the functional requirement is met? Do NOT specify design directives!

6. Correct

Requirements describe in English what the system must do. Thus they must correctly describe that functionality. Only the end-users can make that call.

7. Consistent

Requirements come from various sources. They can be at odds with each other, thus producing inconsistencies between different requirements. These must be resolved early in the requirements process.

8. Complete

Requirements must cover all situations. Of course, that is not possible, but it is a goal that we aim for. Missing requirements means that there are situations for which we have no direction.

A horrible example: Florida pioneered the concept of an electronic bracelet for releasing convicts into the general population years ago. The idea is that you would let the criminal back into society with a smart bracelet around his ankle and if he tried to remove the bracelet, the bracelet would dial a central number and report the break. That happened, the bracelet "phoned" the central number and got a busy signal. There was no requirement for that eventuality so the bracelet "hung" up. The criminal killed an innocent person before being apprehended. Make sure you cover as many cases as you can in getting your requirements as complete as you can.

9. Realistic

Requirements must be reasonable. For example, suppose you were going to publish a recipe and you began with "heat the oven to 5,000 degrees Celsius". Clearly this recipe is not for a conventional oven. Requirements must be obtainable, and thus must be realistic.

10. Traceable

As we have stated, we need to be able to track the requirement back to the stakeholder who defined it. If it comes to pass that the requirement cannot be met, we must inform that stakeholder and assess whether we should cancel the project or not.

5.2.5. Expressing requirements

In most cases, we will be expressing requirements for natural language such as English. Try to remember to make the requirements as precise as you can, using action verbs such as "shall perform, do, design, modify, direct, support". Use precise simple verbs that are clear and unambiguous in describing each requirement. You want to state them as precisely and as simply as you can to avoid ambiguity. Review each requirement with your customer and if you possibly can, build a simple prototype to show the customer what the proposed system will look like.

Second, we can divide requirements into two major classes: static and dynamic. Static clearly means that the requirement is invariant through the life cycle of the project. Dynamic means that the requirement is time sensitive and sensitive not just to time but to certain inputs occurring or not occurring. Static requirements can be described in English of course. If there is a mathematical relationship between data, we can also use recurrence relationships or axiomatic definitions to describe the relationship. For data we could use formal computer science techniques such as Backus–Naur Form (BNF) or object diagrams. For dynamic requirements, we could use artifacts such as decision tables, state diagrams, or transition tables. All of these are computer science techniques for describing dynamic relationships. We do not have time to go into the details here but a standard computer science text will show you the necessary details.

5.2.6. IEEE 830–1998 (Software) requirements specifications standard

In building the final requirements document, it is useful to have a standard with templates to make sure that you get all the different components inserted properly. Fortunately for us, the IEEE (the Institute for Electronic and Electrical Engineers) defined, a couple decades ago, the Software Requirements Specifications Standard (SRS) [5], which is quite useful in serving this purpose. Do not let the word "software" scare you. The SRS is generic for all of Engineering. Let us take a look at what they have done to see how we might use it. Look at the suggested table of contents as shown below.

Table of Contents

1. Introduction
1.1. Purpose
1.2. Scope
1.3. Definitions, acronyms, and abbreviations
1.4. References
1.5. Overview
2. Overall description
2.1. Product perspective
2.2. Product functions
2.3. User characteristics
2.4. Constraints
2.5. Assumptions and dependencies
3. Specific requirements
3.1. External interfaces
3.1.1. User interfaces
3.1.2. Hardware interfaces
3.1.3. Software interfaces
3.1.4. Communications interfaces
3.2. Functional requirements
3.2.1. Mode 1
3.2.1.1. Functional requirement 1.1
. . .
3.2.1.n. Functional requirement 1.n
3.2.2. Mode 2
. . .
3.2.m. Mode m
3.2.m.1. Functional requirement m.1
. . .
3.2.m.n. Functional requirement m.n
3.3. Performance requirements
3.4. Design constraints
3.4.1. Standards compliance
3.5. Software system attributes
3.5.1. Reliability

- 3.5.2. Availability
- 3.5.3. Security
- 3.5.4. Maintainability
- 3.5.5. Portability

3.6. Other requirements

- 3.6.1. Usability
- 3.6.2. Supportability
- 3.6.3. Online Documentation/Help
- 3.6.4. Purchased Components
- 3.6.5. Licenses
- 3.6.6. Legal and Copyright
- 3.6.7. Applicable Standards

4. Supporting information
   - a. Table of contents
   - b. Index
   - c. Appendices

The first two parts of the SRS are basically bookkeeping and introductory material. The first section is the introduction and has five subsections. Section 1.1 describes the purpose of the the project covered by this SRS. It specifies particularly who the intended audience is for this SRS. Section 1.2, the scope section, identifies the product that we are going to produce with this project. For example, the SPENTFUEL project is going to produce a piece of software that will locate the storage locations of all spent uranium fuel of nuclear agencies in North America. It will catalog the precise locations of these in case some regulatory agency needs to go and collect them for eventual permanent disposal. It also describes in the entire contractual environment of the product, where it is physically located for example, on a server, how it can be accessed and including its benefits, objectives, goals and anything else you think might be useful.

Section 1.3 is really important. This is the definitions, acronyms, and abbreviations subsection. Sprinkled all in project management, are many abbreviations which we will often use without thinking of expanding them to our careful readers. The expansions go here, along with abbreviations or specialized definitions that will be used in the SRS. Section 1.4 collects any and all references cited in this document. Of course, the first reference in the section is the reference to the IEEE SRS standard 830–1998; the IEEE recommended practice for Software Requirements and Specification. Each referenced document needs to be identified by title, web address, and the likelihood that you can actually access the web address notated. You also want to specify the sources, where they came from and if appropriate, why they were generated in the first place. Finally in Section 1.5, we want to describe what the rest of this particular SRS contains.

Section 2 outlines the overall description of the product that we are going to create. Each of these subsections are high-level descriptions; the specific requirements will be detailed in Section 3. Diagrams are very useful here.

Section 2.1 is the product perspective. How does this product fit in the scheme of things? Is it self-contained? Is it integrated as a subcomponent of other larger structures? If you can supply any diagrams illustrating its location within the greater structure, that would be extremely useful. How is the design of the final product constrained in terms of the product constraints? Product constraints are considerations such as:

- regulatory policies
- software interfaces
- hardware interfaces
- communication interfaces
- implementation language
- database security
- operating environment

Each one of those issues needs to be considered and described as to how they are going to constrain the construction of the product. For example, if there is a regulation in place specifying that the product must agree to this or that regulation, detail the regulation and its dictates. If you have to write it in a particular programming language such as Java or C#, say so.

Section 2.2 describes product functions. It is basically a description of the major functions that the product ought to perform. Try to organize them in a logical manner dictated by the actual product itself. Again if you can use graphical methods, do so as they are a very clear way of expressing your intent.

Section 2.3 describes the user characteristics of the person actually using this product. What educational level do you expect that person to have? What experience should he have? What is the technical expertise he must possess? Describe any characteristics and skills in the actual usage that are obviously useful.

Section 2.4 covers any constraints we can think of that limits your options in developing the product such as hardware limitations, the necessity of interfacing to other applications, possible parallel operations, reliability requirements, cost constraints, and anything that is going to constrain the actual product itself and its construction.

Finally, Section 2.5 talks about assumptions and dependencies, any factors that are basically underlying the requirements. For example, in writing a piece of software, there is an underlying assumption that the particular operating system on which the application is to be mounted has to function correctly and not fail.

3. Specific requirements

Section 3 is the meat of the SRS and contains the actual requirements themselves. As we will see a bit further, there are eight different ways in which we can organize the material. The obvious way is to structure it by function but we can also organize it in other ways, such as according to stimuli, objects, etc.

To make it specific, suppose we use SPENTFUEL as an example. The numbering here can be used for the numbering of the requirement, for example for entering them into the Requirements Traceability Matrix.

3.1. External interfaces

Think of the application as a collection of interacting artifacts. Now draw a circle around the entire set. Outside that circle are the external artifacts that we need to communicate with. These artifacts are fixed in the sense that their characteristics are interfaces, communications interfaces, and indeed anything that our interior artifacts need to communicate with the outside world. Inside that circle are the relationships between the interacting parts that we need to define by means of our requirements. We make a distinction between the two; external requirements and internal requirements.

3.1.1. User interfaces

User interfaces describe the classes of people that are going to be using the system. It is easiest to define general user classes and then describe the requirements necessary for those folks in each class to interact with the SPENTFUEL system. For example, the site engineer, the person recording the information of where the spent fuel has been buried, would upload reports to the SPENTFUEL application. Every now and then, the SPENTFUEL application would interrogate the site engineer to see if she had anything new to add. Let us list at least four types of people that we have in mind.

- Site engineer
- Regulator
- Database administrator
- Management official

Each of these general classes of people would have specific reports that they would request or deliver, and also respond to specific requests from the SPENTFUEL application. These need to be detailed as clearly as we can.

3.1.2. Hardware interfaces

Hardware interfaces would include any specific hardware that we need to run our application. We do not have any specific requirements as we are assuming that our application is running in a Windows environment using TCP/IP.

3.1.3. Software interfaces

Software interfaces include any specific software applications that we need to depend upon. In this case, the SPENTFUEL application is self-contained.

3.1.4. Communications interfaces

Our basic communication requirement is that we support the TCP/IP protocols both from the external application attaching to the SPENTFUEL application and the SPENTFUEL application itself, communicating with the external users.

3.2. Functional requirements

This particular form of the SRS organizes requirements in terms of the functional requirements required by different modes. Different classes of users would interact with the system in different ways. For example, the site engineer would be concerned only with uploading the reports of his particular nuclear site. The regulator and the management official would be concerned with summary reports. The database administrator would be concerned with detailed reports such as the consistency of the internal tables.

3.2.1. Mode 1 Site Engineer

For mode one, we are concerned about the requirements the site engineer would have for uploading her data and receiving confirmation that indeed the upload had been accepted by SPENTFUEL. We shall now list the functional requirements associated with mode one as we have described previously.

3.2.1.1. Functional requirement 1.1

...

3.2.1.n. Functional requirement 1.n

3.2.2. Mode 2 Inquirers

Mode two would be general inquirers that would like summarized versions depending on whether the requestors are regulators or management personnel. Then follow the requirements for the requestors.

3.2.2.1. Functional requirement 2.1

...

3.2.2.n. Functional requirement 2.n

Thus we continue for the other modalities.

3.3. Performance requirements

The system's performance characteristics are outlined in this section. Include requirements such as:

- specific response time, such as the expected response time for a transaction (best-case, average, worst-case),
- capacity, for example the number of customer transactions the system can accommodate per unit time, and
- degradation modes — what is the acceptable mode of operation when the system is degraded in some manner? This is particularly important in web applications, which is an area that we very rarely think about. What is the web application going to do if it is inundated with far too many requests?

3.4. Design constraints

This section identifies any design constraints on the system being built. Design constraints represent design decisions that have been mandated by the customer and must be adhered to. Some examples include applications, software languages, process requirements, prescribed use of development tools, architectural and design constraints, purchased components, class libraries, and so on.

3.4.1. Standards compliance

If we are following any particular standards, we want to indicate them here. For example, there are IEEE standards on software design that can be brought into play.

3.5. (Software) system attributes

Here, we systematically examine attributes that the system must possess in order to work in an appropriate fashion. Note that we bracket software because the IEEE 830 standard is specifically designed for software standards and as we mentioned before, the ideas contained there can be applied to any system, not just software systems.

3.5.1. Reliability

Requirements for reliability of the system should be specified here. Some metrics that we can use include maintenance access, degraded mode of operation, Mean-Time-To-Repair (MTTR): how long is the system allowed to be out of operation after it has failed? Acceptable bugs or defect rates, categorized in terms of minor, significant, or critical bugs need to be defined. The requirements must define what is meant by a critical defect; for example, the complete loss of data or a complete inability to use certain parts of the system's functionality.

3.5.2. Availability

Specify the percentage of time the system has to be available in terms of hours of use.

3.5.3. Security

We also need to specify the level of security that we demand on our data. How secure does the data have to be, particularly if we permit Internet access to the application?

3.5.4. Maintainability

We need to list the requirements for maintainability. How do we intend to upgrade the system to fix bugs, add features, and so on? Are we going to send out a CD with changes on them or can we do it online directly and remotely? We need to specify the requirements of our maintenance process.

3.5.5. Portability

If we are concerned about portability to other platforms, we need to specify the requirements necessary to make that happen. When are we going to be Linux compliant, Windows compliant, and so on?

3.6 Other requirements

This section lists other requirements that can be useful, and in fact a good idea is to include other requirements that are not covered by the standards here.

3.6.1. Usability

This section includes requirements that affect usability. For example, specify the required training time for a normal user to become productive. Specify the required training time for a power user to become productive in particular operations. Specify measurable execution times for typical tasks or base the new system's usability requirements on other systems that the users know and like.

3.6.2. Supportability

How do we intend to support the system once it gets into the hands of the customer? We need requirements to specify this and this ties into our maintenance activities listed above.

3.6.3. Online Documentation/Help

Describe requirements, if any, for online user documentation, help systems, help about notices, and so on.

3.6.4. Purchased Components

This section describes any purchased components to be used with the system, any applicable licensing or usage restrictions, and anything associated with compatibility and interoperability or interface standards.

3.6.5. Licenses

This section defines any licensing enforcement requirements or other usage restriction requirements that are to be exhibited by the system.

3.6.6. Legal and Copyright

This section describes any necessary legal disclaimers, warranties, copyright notices, patent notices, trademark, or logo compliance issues for any part of the system.

3.6.7. Applicable Standards

This section describes by explicit reference, any applicable standards, with a pointer to the specific sections of any such standards, which apply to the system being described. For example, this could include legal, quality, and regulatory standards; industry standards for usability; interoperability; internationalization; operating system compliance; and so on.

Note that the standard defines six other forms of standard SRS.

5.2.7. Requirements segue into scope

Now that we have constructed the RSD, we can define the scope.

5.3. Define Scope

The inputs for scope definition are the project charter, RSD, and organizational process assets. We have already covered the project charter: it is a high-level view of the project. RSD: we have just covered that; it details the precise requirements that the product has to possess. Finally, the organizational process assets are things that the organization will bring to bear on the project itself. We want to make sure that our project is aligned with the organizational business goals. The organization may have formal or informal policies, procedures and guidelines that structure scope management. The organizational policies and procedures, as they pertain to scope definition management, are probably recorded in templates in the organization. Any historical information we can get from previous projects that are similar to this would be highly useful. This is then the beginning of the definition of our scope.

The tools and techniques that we can use for scope definition are shown in the list below.

1. expert judgement
2. product analysis

3. alternatives definition
4. facilitated workshops

Expert judgement means getting expert opinions from SMEs. These people can come from other functional units within the organization. They could be consultants; they could be particular stakeholders; they could come from professional and technical associations or industry groups; they could be just people in general, who know something about the product that we are trying to build at this point.

The second tool; product analysis, is a technique to obtain a better understanding of the product that the project is going to create. There are techniques from manufacturing engineering that we can use, such as product breakdown analysis, systems engineering, value engineering, value analysis, functional analysis, and quality function deployment. Any one of these tools is quite useful, and while we do not have time to describe them in detail, they could be checked out for profitable use.

The third tool we can use is alternatives definition. The purpose of the project is to create a product that can solve a particular problem or generate a certain revenue stream. Are there other ways of doing this than just running a project? For example, it may be preferable to outsource the entire project to China, if that makes economic sense. It may be preferable not to do the project at all. In doing alternatives definition, you want to use tools like brainstorming, lateral thinking, storytelling, and scenario development; techniques that get a group together and have people blue-sky about what the problem is and what alternatives might lead to the same solution.

The final tool we can use are facilitated workshops. These are focused structured sessions of particular stakeholders who try to map the projects or expectations into the project scope and plan activities.

All of these ideas try to flesh out the actual scope of the project itself. Now we have mapped the input of the three major sources, using the four tools and their subsets, to the output of the Define Scope phase, which is going to be the Project Scope Statement (PSS). This is an important document because it details the scope of the entire project; that is, the steps we need to take to ensure the appropriate quality of the final product. There will also be updates to other project documents, such as the Project Management Plan, the Risk Plan, the Quality Plan, and so on.

The output of Define Scope is the PSS document. This is the detailed description of the project's deliverables and the work required to create those deliverables. This work is defined in terms of project management steps that we have to execute the project. The document provides a common view of the project that ALL of the Stakeholders can see. It describes the project's objectives. It defines the boundary between what is In-Scope and what is Out-of-Scope.

The PSS will include or at least point to, the following items:

1. Project objectives
2. Product requirements description
3. Project requirements
4. Project boundaries
5. Project deliverables
6 Product acceptance criteria
7. Project constraints
8. Project assumptions
9. Initial project organization
10. Defined risks
11. Schedule milestones
12. Fund limitations
13. Cost estimates
14. Project CC/ICC requirements
15. Project specifications
16. Approval requirements

We now expand on some more detail of each one of these 16 points.

1. Project objectives

The PMBOK defines objectives as being the *measurable success criteria of the project*. The key word in this definition is measurable; we must be able to assign the metric or number to that particular objective. Objectives can come from many areas such as cost, schedule, or quality, and in fact, objectives normally have these three attributes associated with them. One of the ways to characterize objectives is to use SMART, i.e., Specific, Measurable, Agreed to, Realistic, and Time-constrained, as described previously in Chapter 3.

Some typical examples of objectives for projects in general might include:

- benefits of the project (ROI justification)
- operational improvements
- enhanced readiness
- productivity improvement
- market opportunities
- improved customer service

2. Product requirements description

The product requirements description precisely defines the major characteristics that the product must have, which is a large component of the RSD we have just defined at the beginning of this chapter.

3. Project requirements

The project requirements specified the conditions the project itself must meet to satisfy certain conditions, such as a contract, standard or specification. The needs of the stakeholders, their wants and expectations have to be analyzed and mapped into prioritized project requirements.

4. Project boundaries

One of the most important questions in the scope definition is the question "is it in scope or not?" It is vitally important that we understand what the project boundaries are and what we are supposed to be building and more importantly, what we are not supposed to be building. This is important for Integrated Change

Management and project rebaselining if we have to do that. We want to EXPLICITLY specify items that will not be done in case of any possible ambiguity by somebody outside the project making the assumption that we are going to do something that we have decided not to do.

5. Project deliverables

This section precisely defines the project deliverables. We make a distinction between in-project and out-of-project deliverables. Project deliverables include elements like project management reports, Earned Value Management (EVM) reports, risk assessments, and so on. Out-of-project deliverables include things that are delivered externally to the project such as the RSD, the project plan, and the quality plan.

6. Product acceptance criteria

This is a very important section for each product and indeed, each product deliverable. The PM must specify the conditions under which the deliverable has to be accepted. The major reason for this is that the customer does not want to accept the product unless he is guaranteed it functions in the way specified in the RSD.

7. Project constraints

This section elaborates on the constraints initially touched upon in the project charter. Anything that is within the scope that can constrain the project needs to be stated here. For example, a phase may have a predefined budget for completion such as requirements. There may be predefined timelines that have to be honoured. One can imagine the Y2K timeline, for example, as being a very strong project constraint in upgrading software to work into the new millennium.

8. Project assumptions

Projects assumptions are all over the place and they are mostly invisible. Here we want to detail as many as we can. For example, we assume that there will be adequate funds released by management in time to complete each sequential phase. If we are concerned about the price of oil, then our stated assumption is that the price of oil would be constant at $90 a barrel. Perhaps if we are concerned about the price of the Canadian dollar, we can assume it will remain within 5% of the American dollar.

9. Initial project organization

In this section, we name the core team members of the project and the major stakeholders. An organizational chart is a good tool to define the relationship between these people, and it is important that we spell out the names of the people that we have as team members and stakeholders and how we can get hold of them in case we need to.

10. Defined risks

We started our risk identification in the preparation of the charter. Here we continue to elaborate the risks as we dig further into the definition of the project. The definition of the scope will identify additional risks that we want to detail. We start out for the definition of the Risk Registry (more on this later in Chapter 11) and plan on these successive elaborations of risks through our planning processes.

11. Schedule milestones

As with risk, we have already defined major schedule milestones in the Charter. The PSS will define more now that we understand the process better. List the important milestone dates. Are there imposed dates that we have to meet? State them. As we understand the project better, more delivery dates will be defined, in a similar process to the risk elaboration.

12. Fund limitations

Specify any money limitations that exist either on a phase basis or in total value. Specify also the amount of the contingency funds allocated to the project. Remember contingency funds refer to an overflow money buffer that we can access if we underestimate the cost of a particular phase. In software projects, the project contingency fund is typically set at 30% of the total value of the project. In other areas, those tolerances may be much tighter.

13. Cost estimates

Estimate the total project cost that we can determine at this point and give an indication of the accuracy of that estimate. A common way to do the estimate is to use a three-point estimate: worst case estimate, most likely estimate, and best case estimate. The rule of six says that you take the most likely estimate multiplied by four, add to it the worst case estimate and the best case estimate and divide the whole by six, to give you a more representative estimate than just a simple average.

14. Project CC/ICC requirements

Configuration Control (CC) and Integrated Change Control (ICC) refer to the way we are going to perform change management on physical and structural elements of the project. We want to specify how we are going to do that, the level of detail required to perform CC/ICC, and anything else that can be related to change management in general. Note that configuration control can be considered as a subset of integrated change control but for historical reasons, we normally separate them out.

15. Project specifications

In this section, we identify the specification documents with which the project must comply. For example, it might follow the IEEE 1059 project management plan standard. Any other standards that we are going to use in the project should be detailed here. As a second example, we might make a requirements reference to the IEEE 830–1988 SRS.

16. Approval requirements

Last, but not least, we want to identity specifically, by name and location, the people who will approve the named deliverables and the conditions under which they MUST be accepted.

The second major output to be created in this phase, are project document updates. As we expand the PSS, changes may be required to other project documents, such as the:

- project management plan
- RSD
- requirements traceability matrix
- stakeholder register
- risk register (see Chapter 11)

As we do this, these will all have to go through integrated change control, as these documents will have been baselined by this point.

5.4. Create Work Breakdown Structure

The Scope Management Plan also documents how the Work Breakdown Structure (WBS) will be created and defined. The WBS is critical to project success [6]. It subdivides major deliverables into smaller and more manageable components. Formally speaking, the WBS is a deliverable-oriented hierarchical description of tangible project components that organizes and defines the total scope of the project. Note the word "total". Everything that will be done in the project must be represented by the deliverable-oriented WBS. The WBS is the skeleton on which all the other project information is hung.

The WBS provides the foundation for integrating all of the work package details and deliverables with all the other aspects of the project including initiation, planning, execution, monitoring and controlling, closing and risk management. The benefits of a deliverable-oriented WBS are many. The first benefit is that we have better communications with our sponsors, stakeholders and team members. We have much more accurate estimations of tasks that have to be done, the risks in doing those tasks, the costs associated with them, the scheduling of the tasks and best of all, we have an increased confidence that we have covered all the work necessary to complete the project, ensuring that nothing is left out. The WBS supports clarity in the project. It decomposes the project scope into deliverables and in fact, each individual work package is characterized by the fact that each creates at least one deliverable. It supports the definition of the work effort required for effective management and will drive the controlling mechanism to ensure the project is on time and on budget. (We shall see that in Chapter 7 when we talk about Earned Value.) The creation of the WBS defines the scope in terms of deliverables that all the stakeholders can understand and it relates the work packages to the Organizational Breakdown Structure and the Responsibility Assignment Matrix, as we will see in later chapters.

To create the WBS, we need two things. The first is the general topology of the WBS. The second is the description of each of the individual work packages that constitute the WBS. We look at the WBS topology first. In defining the topology of the WBS, we can look at it from different points of view. We could take a product-oriented view. We could take a phase-oriented view or a function-oriented view or most likely a technical view. So, for example, with the SPENTFUEL project, the most likely breakdown is to generate WBSs, one for the technical aspects of actually constructing the software itself and the second, from the management aspect which is the project management control of the project. Figures 5.2(a)–(g) and 5.3(a)–(j) illustrate possible breakdowns for each of these WBSs. See [6] for more details.

The second part of the WBS creation is the decomposition of the work into the work packages — the lowest level WBS components. We have to subdivide the project deliverables into smaller, more manageable components. This work is much like design: when we have a high-level view of the product we want to design, we break it into smaller components until we can clearly see the work that needs to be done. That is what we do here: we continue the breakdown until we expose enough structure that we can estimate the cost of work and see how many resources we need to apply to get the work done. Another criterion we can use is that we continue

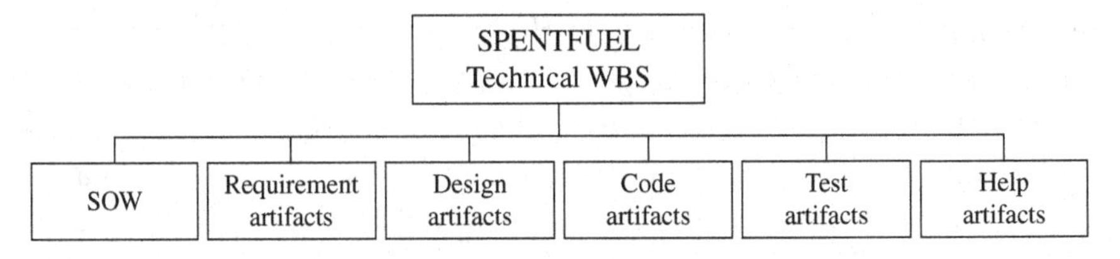

Figure 5.2(a). SPENTFUEL technical work breakdown structure.

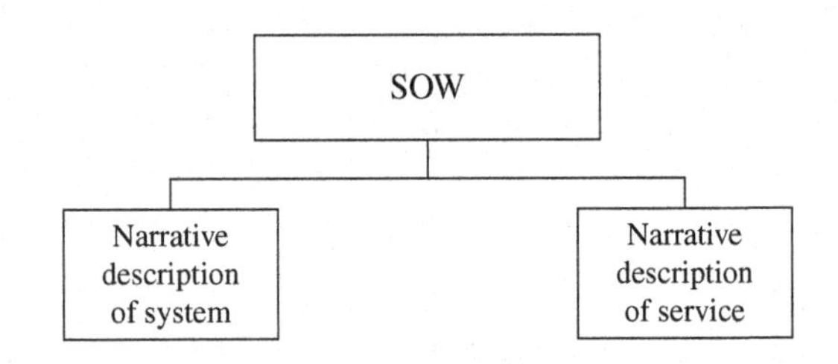

Figure 5.2(b). SPENTFUEL technical WBS — SOW.

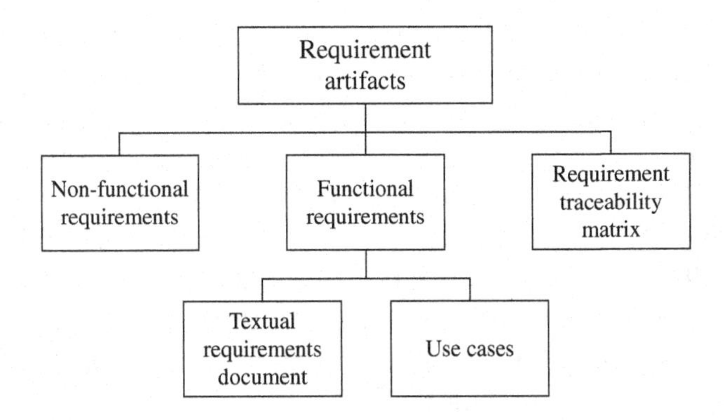

Figure 5.2(c). SPENTFUEL technical WBS — requirement artifacts.

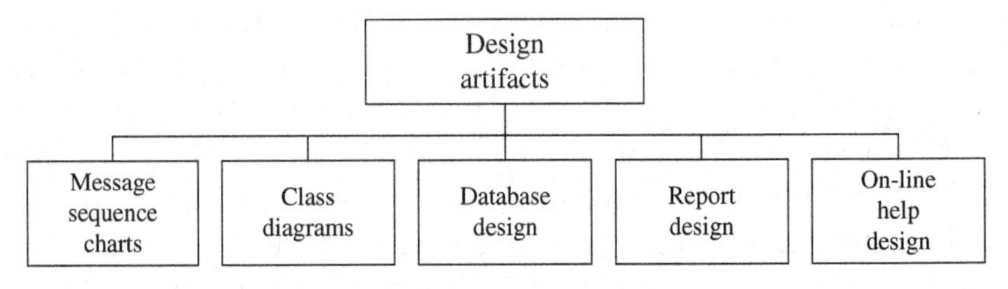

Figure 5.2(d). SPENTFUEL technical WBS — design artifacts.

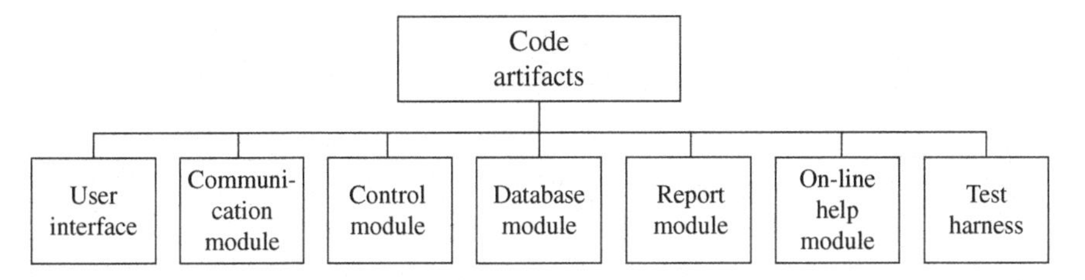

Figure 5.2(e). SPENTFUEL technical WBS — code artifacts.

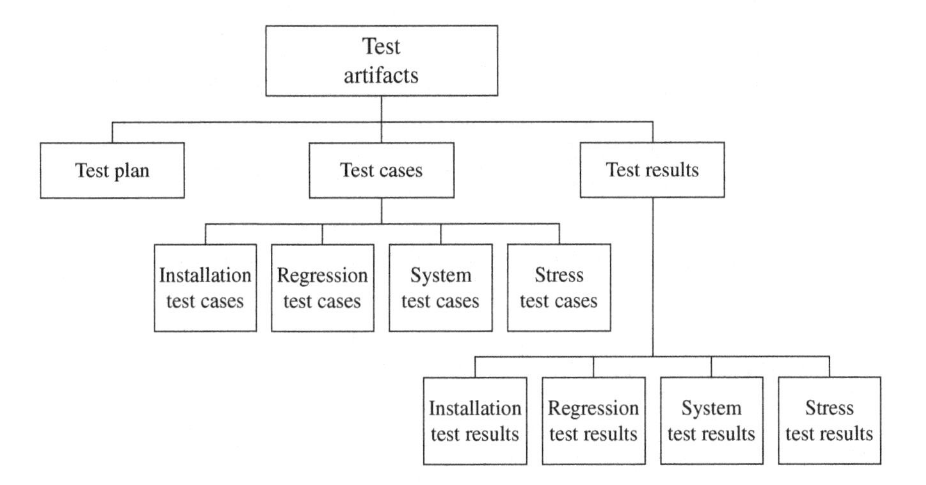

Figure 5.2(f). SPENTFUEL technical WBS — test artifacts.

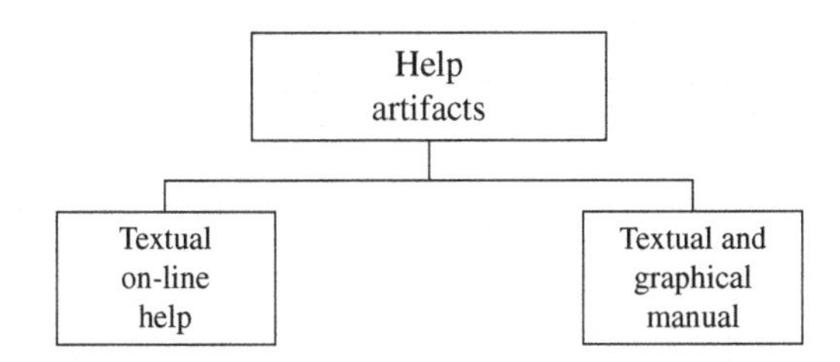

Figure 5.2(g). SPENTFUEL technical WBS — help artifacts.

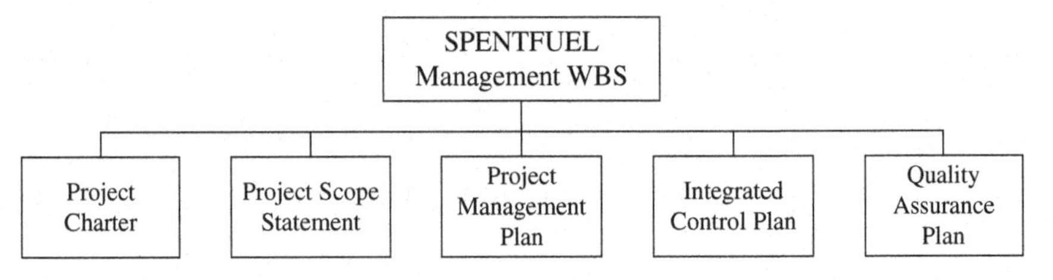

Figure 5.3(a). SPENTFUEL management work breakdown structure.

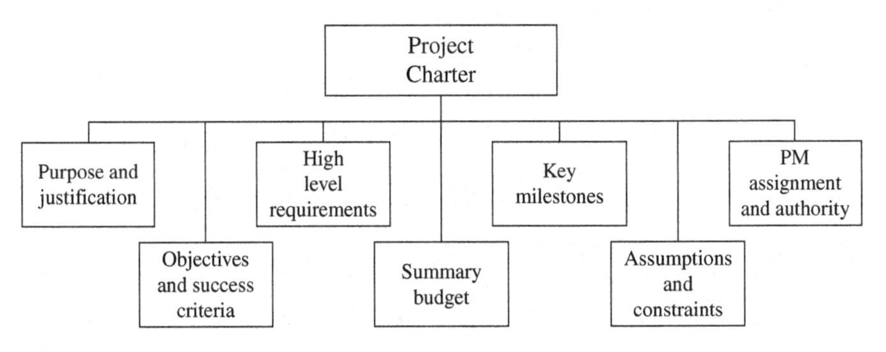

Figure 5.3(b). SPENTFUEL management WBS — project charter.

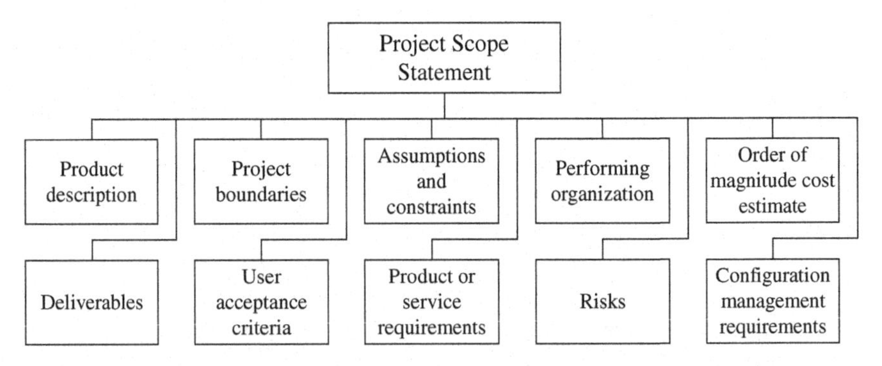

Figure 5.3(c). SPENTFUEL management WBS — project scope statement.

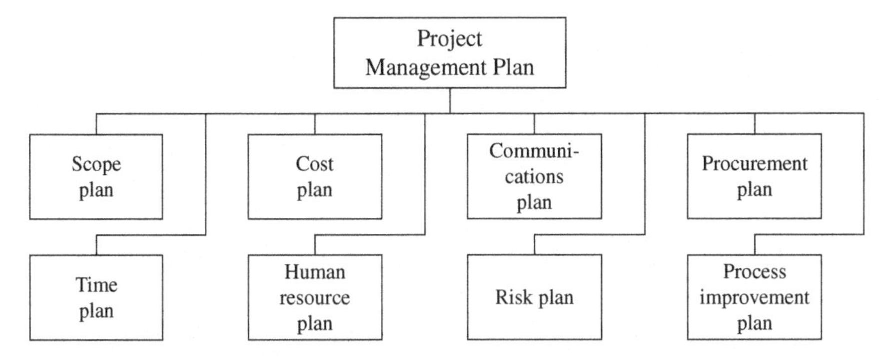

Figure 5.3(d). SPENTFUEL management WBS — project management plan.

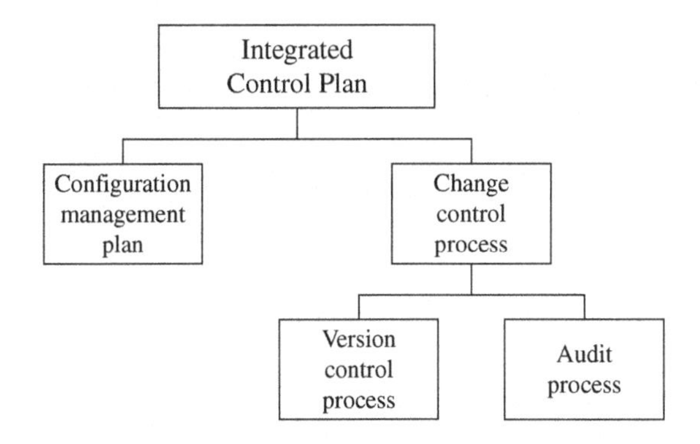

Figure 5.3(e). SPENTFUEL management WBS — integrated control plan.

Figure 5.3(f). SPENTFUEL management WBS — quality management plan.

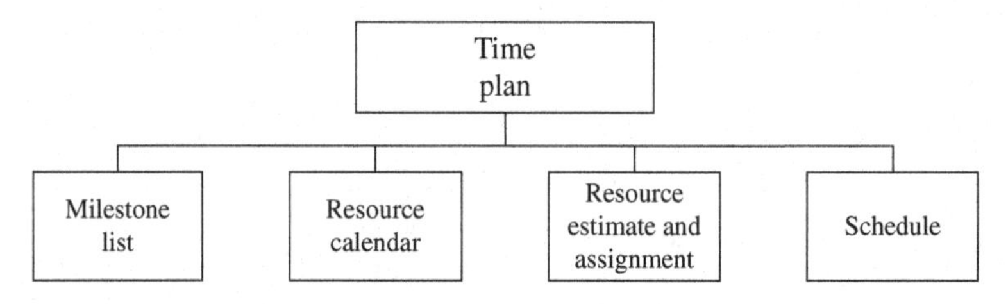

Figure 5.3(g). SPENTFUEL project management plan WBS — time plan.

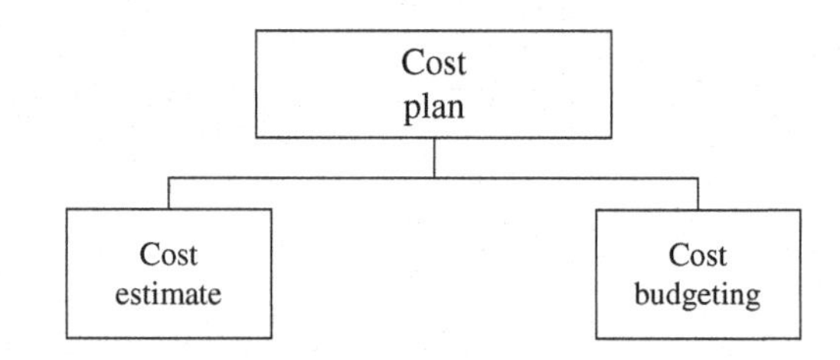

Figure 5.3(h). SPENTFUEL project management plan WBS — cost plan.

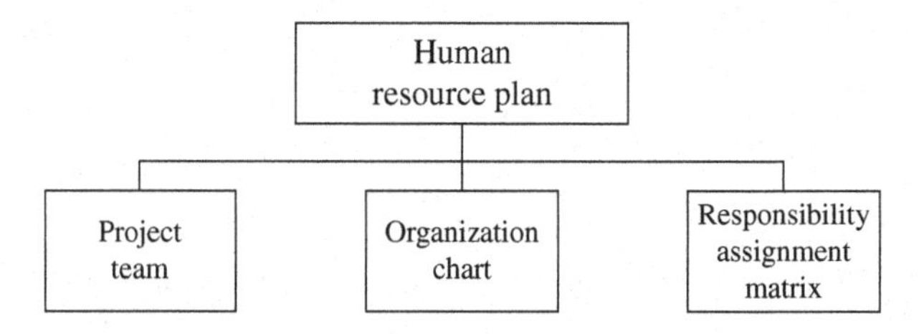

Figure 5.3(i). SPENTFUEL project management plan WBS — human resource plan.

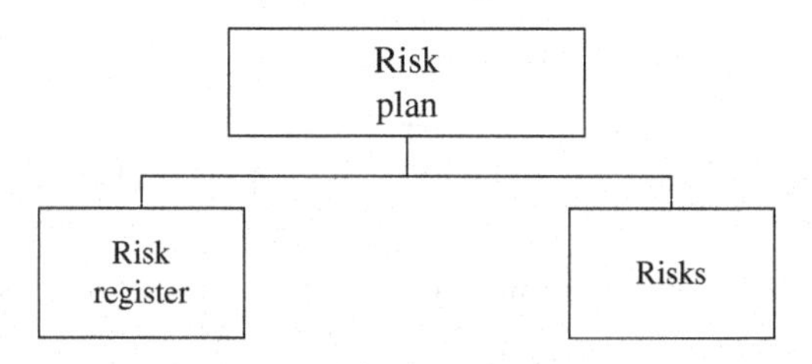

Figure 5.3(j). SPENTFUEL project management plan WBS — risk plan.

the decomposition until we can see the risk exposure to not completing that particular work package.

We can take a look at the decomposition as a four-step process. Step one is to identify the major deliverables of the projects. Step two: can we estimate the cost and times for the deliverable at this point? If so, go to Step four. Step three: If we cannot break that particular work package into its proper subcomponents, go back to Step two. Step four is to verify the correctness of the decomposition. By the time we are finished, we will have broken everything down into executable components. The industrial rule-of-thumb follows the 80-hour rule, whereby WBS continues to be hierarchically decomposed until the lowest level deliverables (work packages) are less than 80 hours (roughly two weeks of work).

The companion to the WBS diagram is the WBS dictionary. It contains a description of every rectangle that we have defined in the WBS. Each dictionary entry has to have at least six criteria stated, and we list them now.

1. Status and completion can be measured.
2. We have clearly defined start and end events.
3. The work package has a deliverable which we name.
4. Time and costs can be accurately estimated.
5. The work package duration is within acceptable project limits.
6. The work assignments are independent.

In addition to the six criteria, we also need to list the name of the person that will be the sign-off acceptor of the deliverable of this work package. Anything else that is of use in defining the project management plan can be included in the description of the work package.

The creation of the deliverable of a work package will normally require the execution of several distinct activities. Activities are a subset of the work package. They are pieces of work that have to be done but do not necessarily generate an external deliverable. As a rule of thumb, WBS and work packages have descriptions that are nouns, whereas schedule activities have descriptions that start with verbs. For example, "remove" debris, "pour" concrete, "design" class diagrams, "conduct" user acceptance test are schedule activities. We defer further discussion to Chapter 6.

In concluding this section, it is instructive to take a look at the consequences of having a bad or nonexistent WBS. You can have incomplete project definition, leading to project extensions and cost overruns. Unclear work assignments will result if we do not understand completely the work that has to be done. Without a WBS, it is highly likely that we will have scope creep and massive scope changes. We are almost guaranteed to have budget overruns, missed deadlines, and timeline slippages. We are likely to create an unusable product and fail to deliver some of the elements of the project scope because we have not actually detailed all the work that needs to be done.

The WBS is incredibly important to get right, because everything else that we do in the project is dependent on the accuracy of that structure.

5.5. Validate Scope

When the Project Scope Statement (PSS) is completed, it must be validated by the appropriate person for sign-off. Before the PSS can be put under ICC, the project manager must prove to the person for sign-off that the decomposition is both necessary and sufficient. You must have the approval of all the stakeholders that this is the case. If the project is terminated early, the scope statement is going to determine the level of completion, which is useful in the legal wrapping up of the proceedings. Note that this scope validation is concerned with the acceptance of the work, not the correctness of the work. The correctness of the work is the job of quality control. Normally, both are done in parallel to make sure that both goals are achieved. The tools and techniques for scope validation are inspections. It can have other names as we shall see later in the book, such as reviews, walk-throughs, audits, product reviews, etc. Finally, the most important thing from the scope validation is to get sign-off on the actual PSS itself. Any defects that have been discovered in the walk-through of the validation need to be documented for insertion into the lessons learned document that we will write at the completion of the project.

5.6. Control Scope

Scope change control must be implemented from inception through completion of the project. Change requestors or focal points need to be designated. A standard form should be used for change control and the process for submitting a change request needs to be communicated so that people are aware of it. The turnaround time needs to be set and all approved or unapproved requests should be documented. Requests should be prioritized and certain defined categories of requests can be approved automatically. In case of emergency, approval without review may be granted.

Scope control is concerned with influencing the factors that can cause scope changes, and if the changes do happen, to make sure that they are done in a structured way, that all the stakeholders understand the need for the changes and approve of the changes. Once approved by the proper authorities, scope control manages the actual changes to make sure they are done in an organized manner. Scope control has to be integrated with other control processes such as schedule, cost, quality, and risk. Now you might think: is this not just another example of ICC? Yes, you are right, but this is so critical that the ICC for scope has to be explicitly exposed just as we had to expose the ICC for the project plan. Scope changes can be initiated through the normal ICC request form procedure. Scope changes can also be transmitted orally, directly or indirectly, internally or externally initiated, or legally mandated. Most scope changes are a result of unforeseeable factors such as external events, changes in government regulation, etc. In addition, they can be caused by errors or omissions in product requirements that we missed in writing the original RSD. They can be caused by errors or omissions in the scope definition that we missed; they can be value-added changes that are caused by technology changes. For example, a project failed because the industry changed from a glass TTY environment to a Windows environment and the project failed to change scope to take that into consideration. Often scope changes may be initiated by risks eventuating that cause us to change direction.

There are four possible outputs from Control Scope. First, the scope can be changed. Any approved changes have to go through the ICC cycle since the scope document has been baselined. It is important to note that approved changes will very likely increase the cost of the project and lengthen the time for completion. Second, corrective action may be ordered. We need to document any activities necessary to bring the future work in line with the original plan. Thirdly, there may be some lessons learned in the execution of the project, and it is important that we record these lessons learned for the document we are going to write at the end of project completion. Fourthly, scope changes may force us to adjust the baseline. This means resetting the project management plan and any other plans impacted by the approved scope changes.

5.7. WBS Exercise

From the moment you get up from bed every morning until you arrive at school or at work, capture what you do in a WBS with no more than 3 layers and no more than 20 work packages.

5.8. Multiple Choice Questions

1. A key activity for achieving customer satisfaction is to define:
   (a) The business case
   (b) Requirements
   (c) Product specification
   (d) Change control

2. What is the bottom level of a work breakdown structure known as?
   (a) Work packages
   (b) Milestones

(c) Project components
(d) Deliverables

3. During project scope planning, the work breakdown structure should be developed to:
   (a) The sub-project level
   (b) The level determined by the PMO
   (c) A level allowing for adequate estimates
   (d) The cost center level

4. The decomposition process is a technique used to construct a:
   (a) Precedence diagram
   (b) Critical path method diagram
   (c) Variance analysis
   (d) Work breakdown structure

5. A precise description of a deliverable is called a:
   (a) Specification
   (b) Baseline
   (c) Work package
   (d) Work breakdown structure

6. Your company has to offer its service on the internet to increase its market share. You are asked to start planning for this project. What is your first step?
   (a) Identify the risks
   (b) Plan the scope
   (c) Establish a resource plan
   (d) Complete a cost and schedule estimate

7. All of the following assist in determining the impact of a scope change except:
   (a) Project charter
   (b) Baseline
   (c) Performance measurement
   (d) Milestones

8. A requirement document is important because it:
   (a) Provides the basis for making future project decisions
   (b) Provides a brief summary of the project
   (c) Approves the project for the stakeholders
   (d) Provides criteria for measuring project cost

9. Reviewing work products and results to ensure that all were completed satisfactorily and formally accepted is to:
   (a) Manage risk
   (b) Control quality
   (c) Control change
   (d) Validate scope

10. Due to cuts in funding, your project has been terminated. The "validate scope" process:
    (a) Should be delayed until the project is completed
    (b) Should determine the correctness of the work results
    (c) Should establish and document the level and extent of completion
    (d) Will form the basis of the project audit

References

1. Project Management Institute. *A Guide to Project Management Body of Knowledge* (PMBOK), 5th Edn. PMI, 2013, p. 558.
2. Dion, R. Process Improvement and the Corporate Balance Sheet. IEEE Software (July 1993), pp. 28–35.
3. DeMarco, T and Lister, T. *Peopleware: Productive Projects and Teams.* Dorset House, 1984.
4. The CHAOS Report 1995, the Standish Group, Boston, Massachusetts, USA, 1995.
5. IEEE Standard 830–1998. Software Requirements Specification. IEEE, 1998.
6. Project Management Institute. *Practice Standard for Work Breakdown Structures*, 2nd Edn. PMI, 2006.

CHAPTER SIX

Time Management

6.1. The Purpose of Time Management

Now that we have finished the Project Scope Statement, and the corresponding Work Breakdown Structure, we are now in a position to construct a diagram of the time execution of the project. We have to look at each of the work packages and their corresponding activities, assign resources to execute the activities and estimate the time for completion. Then we have to examine the relationship amongst the activities, their interdependencies and then schedule these in a coherent manner to complete the project. That is the purpose of time management and Chapter 6.

6.2. Time Planning

This process details how time-related activities will be managed through the planning, executing, monitoring and controlling, and closing of the project (or phase).

6.2.1. Work packages and activities

The Work Breakdown Structure and the Work Breakdown Structure Dictionary describe the work packages that need to be done to complete the project. Work packages are broken down into smaller components called schedule activities, which are action steps for project execution. Schedule activities have dependencies or logical relationships that prescribe the connections of the activities in a sequential manner from the start to the end of a project. Each project has a single start and single end point. Now we need to logically organize the time sequence of the execution of these activities. To do that, we need to estimate the effort and the number of resources needed to complete each of the activities. That is the purpose of the rest of this section.

For example, recall "Class Diagrams" is a work package (lowest level WBS with less than 80 hours of work) of "Design Artifacts" in the "Technical WBS of SPENTFUEL" in Figure 5.2(d). The activities are decomposed as follows:

- activity 1: derive class diagrams for SPENTFUEL
- activity 2: conduct technical formal review of class diagrams of SPENTFUEL
- activity 3: update class diagrams for SPENTFUEL based on technical formal review
- activity 4: post class diagrams for SPENTFUEL on project web page

The sequence of execution of the above activities will be: activity 1 → activity 2 → activity 3 → activity 4.

6.2.2. Activity-on-arrow diagram

An activity is represented by a circle or node and the connection between an activity and its downstream neighbour is represented by an arrow. We have two choices in designing such a diagram. We can put the activity on the arrow or we can put the activity on the node (or the circle). We refer to this as an Activity-On-Arrow (AOA) or Activity-On-Node (AON). Figure 6.1 gives an example of an AOA Diagram, also known as the Arrow Diagramming Method.

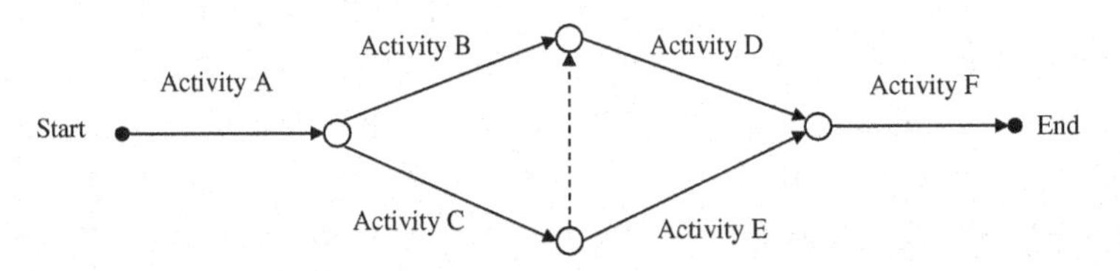

Figure 6.1. Activity-on-arrow diagram example.

Note the single start and single end points in the diagram. Activities are connected at the nodes. The descriptions of the activities (e.g., Activity A, Activity B) are labeled on the arrows. The sequence of execution begins with activity A that has to be completed before activities B and C can start. Note also the dummy activity (dotted arrow) between Activity C and Activity D. It illustrates the sequencing that Activity D cannot start until both Activities B and C have been completed.

6.2.3. Activity-on-node diagram

The same project can be represented by an AON Diagram (see Figure 6.2), also known as the Precedence Diagramming Method, which is more commonly used in the industry.

Note that the descriptions of the activities are labeled in the nodes of rectangular boxes. Activities are linked by precedence relationships. These precedence relationships can be of four kinds. Suppose activity A has an arrow leading downstream to activity B. The four relationships are:

- Finish-to-Start (activity A must finish before activity B can start)
- Finish-to-Finish (activity A must finish before activity B can finish)
- Start-to-Start (activity A must start before activity B can start)
- Start-to-Finish (activity A must start before activity B can finish)

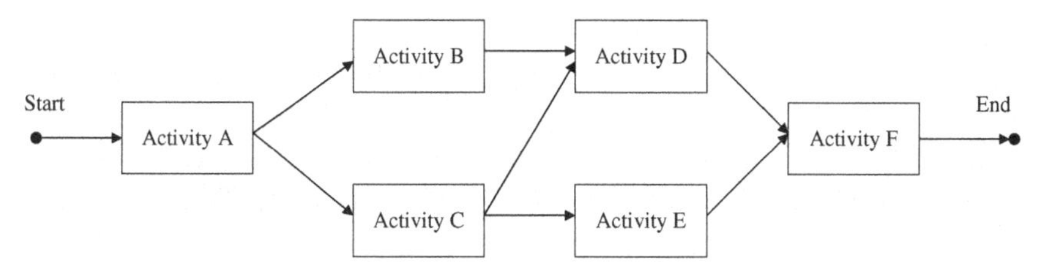

Figure 6.2. Activity-on-node diagram example.

The default precedence relationship is Finish-to-Start, unless otherwise labelled on the arrow (more in Section 6.3.5).

Each node can be described by a template that includes the following information (more in Section 6.3).

- ES Early Start
- EF Early Finish
- Activity number
- Activity description
- Duration
- LS Latest Start
- LF Latest Finish

6.2.4. Rules for diagramming methods

The following rules are generally followed for the diagramming methods.

- single start, single end
- each activity has at least one predecessor and one successor
- no loops
- description of an activity starts with a verb (i.e. an action), for example, "remove" debris, "pour" concrete, "conduct" user acceptance test

Note that either technique does not permit looping in the diagram itself. In reality, we have lots of loops. For example, suppose we have an activity such as "Sign-off on requirements", and we failed that activity. Then we would have to redo the activity which would in effect cause a loop in the diagram. To represent that, we need a completely different type of diagramming technique. A recently developed technique called Graphical Evaluation and Review Technique (GERT), can be used in this case. The details are beyond the scope of this book.

6.2.5. Types of dependencies

There are three types of dependencies:

- External

  An external dependency involves a relationship between project and non-project activities that are beyond the control of the project team. For example, the shipment of hardware from a third-party vendor for laboratory setup is beyond the control of the project manager (PM). Although external dependencies are beyond the control of the project team, it is still necessary to set a deadline for them with an adequate buffer for contingencies. As a SPENTFUEL example, the external dependencies for "activity 1" in Section 6.2.1 can be "object-oriented design training complete" or "CASE tool for drawing class diagrams available".

- Mandatory

  Mandatory dependency is also known as hard logic. Due to nature of the work, an activity must be completed before the next one starts. For example, "remove debris" must be done before "pour asphalt" in constructing a driveway. For the SPENTFUEL example in Section 6.2.1, "activity 1" must be completed before "activity 2" starts.

- Discretionary

  Discretionary dependency is also known as soft logic. The activities have no relationships, yet at the discretion of the PM/team, they are linked up sequentially for ease of understanding and management. For example, a programmer is assigned to five programming tasks. Instead of tackling all five at the same time, she may choose to prioritize the tasks and tackle one or two simultaneously, while the remaining will begin upon completion of the top priority ones. In this case, unrelated tasks are linked according to preference or priority, and the order of execution is prescribed.

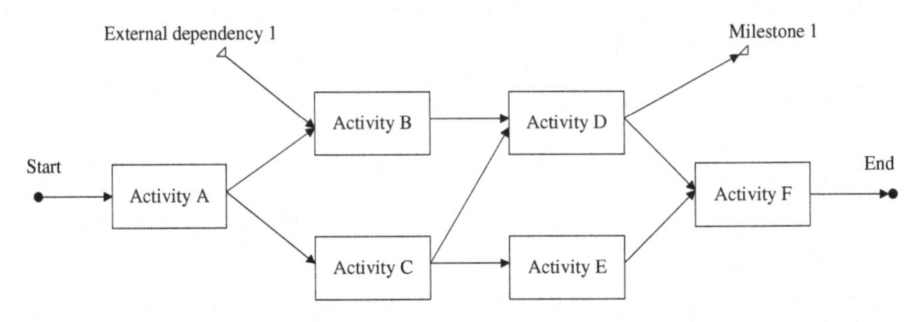

Figure 6.3. External dependencies and milestones example.

6.2.6. Milestones

Milestones mark major accomplishments during project execution when certain activities have been completed. A milestone is typically represented by a diamond in an AON diagram linked by an arrow or arrows flowing out of an activity, or multiple activities. Likewise, an external dependency is also represented by a diamond, yet linked by an arrow or arrows flowing into an activity, or multiple activities. Figure 6.3 shows an AON diagram with external dependencies and milestones.

Note that Activity B cannot start until Activity A is done and External Dependency 1 is satisfied. When Activity D is done, Milestone 1 is accomplished.

Both external dependencies and milestones are markers. They are not activities and are of zero duration. As a SPENTFUEL example, a milestone for "activity 4" in Section 6.2.1 is "class diagrams completed".

6.2.7. Resource assignment and duration estimation

In this phase, the type of resources (e.g., programmer) and quantity of resources (e.g., 4) required to perform the schedule activities are estimated and assigned. A Responsibility Assignment Matrix (RAM) can be developed (see example in Table 6.1) to understand the roles and responsibilities of each person on the schedule activities.

Table 6.1. SPENTFUEL control module RAM example.

	PM	Architect	Designer	Programmer	Tester
Task A	primary	secondary	secondary	secondary	
Task B		primary	secondary		
Task C			primary	secondary	
Task D			secondary	primary	
Task E					primary

Table 6.2. SPENTFUEL control module resource spreadsheet example.

Resources \ Time	1	2	3	4	5	6	7
Resource A	1					1	1
Resource B	0.5	1					
Resource C		1	1	1			
Resource D		1	1	1	0.5		
Resource E	1	1	1	1	1	1	1

Based on the responsibility matrix and AOA/AON diagram, a resource spreadsheet (see example in Table 6.2) can be developed to understand the loading of each resource. The key is not to overload or under-utilize any resource during the period that the resource is assigned to the project.

Note that it is not unusual to share a resource across multiple projects. Once the availability of a resource is agreed amongst PMs, the availability of the resource must be capped at such availability to avoid overloading. For example, if we manage the development of Control Module and Database Module of SPENTFUEL as two projects, and a designer is going to be 50% available for the Control Module and 50% available for the Database Module, then he should only be loaded at 0.5 in the resource spreadsheet for each project.

A series of examples will be used for calculating the duration of a schedule activity. For example, "activity 1" in Section 6.2.1 will take 10 staff-days of effort to complete:

- Number of Resources

  For example, by assigning 2 resources to work on the activity,

$$\text{duration} = \frac{\text{effort}}{\text{number of resources}} = \frac{10}{2} = 5 \text{ days}$$

- Productivity

  Resources are not 100% productive as there will always be administrative overhead. In an IT environment, 70–80% productivity is typically used, depending on the seniority of the resource. Examples of overhead activities include checking non-project emails, attending non-project meetings/trainings, travel for business trips, etc.

  Continuing with the same example, by assigning two resources, each with 80% productivity to the activity,

$$\begin{aligned}\text{duration} &= \frac{\text{effort}}{\text{number of resources} \times \text{productivity}} \\ &= \frac{10}{2 \times 80\%} \\ &= 6.25 \text{ days}\end{aligned}$$

- Availability

  Availability refers to the percentage that the resource is dedicated to the activity.

  Using the same example, by assigning 2 resources with 80% productivity and 50% availability to the activity,

$$\begin{aligned}\text{duration} &= \frac{\text{effort}}{\text{number of resources} \times \text{productivity} \times \text{availability}} \\ &= \frac{10}{2 \times 80\% \times 50\%} \\ &= 12.5 \text{ days}\end{aligned}$$

In summary,

$$\text{duration} = \frac{\text{effort}}{\text{number of resources} \times \text{productivity} \times \text{availability}}$$

Also,

$$\text{cost} = \frac{\text{unit cost } \times \text{effort}}{\text{productivity}}$$

Note that the number of resources and availability have no impact on cost. If unit cost for the above example is $1,000 per day,

$$\text{cost} = \frac{\$1{,}000 \times 10}{80\%} = \$12{,}500$$

6.2.7.1. *Duration estimation*

Duration estimation is perhaps the most complicated and least accurate of all our processes in project management. Recall from our discussion of Chapter 2 that we need two major estimates in project management: the estimate of the size of the work to be done and the estimate of the effort in terms of person-months to do that work. It is also important to specify the confidence levels of our estimates and where we got them from. There are several techniques that we can use to obtain our estimates, namely:

- expert judgement
- analogous estimates
- parametric estimates

For expert judgement, we use the advice of people that have estimated before and who are presumably subject matter experts. In dealing with experts, the Delphi approach is a good way to start. We begin by gathering together a group of experts. We explain the problem and send them away to separate rooms where they come up with estimates on their own without any communication with each other. Then they all meet again and exchange their information in a public forum. Then they go back to their separate rooms.

After three or four cycles of this process, they will normally converge on a uniform answer, which generally speaking, is highly accurate. The second way is to use analogous estimates. If we have done similar projects in the past, we use those values to determine the new estimates. The third way uses parametric estimates. This is dependent on measuring, in real time, the amount of work that we are actually doing in the field. In software engineering, for example, we can use the concept of function points. Function points are way of measuring the complexity of software and we can get industry ratings specifying how many function points per month the average software engineer can accomplish. If you have such parametric estimates available, by all means use them. This is field dependent. In some fields of engineering, standard tables are published indicating the average estimates of time needed to actually do the work.

6.2.7.2. *An example of parametric estimation: function points*

Function Points are a measurement of the complexity of software and can be used both to estimate the size of the work and productivity rates to do the work. If the distance from London (Ontario) to Toronto is 200 km and I can drive at 100 km/hr, then I can quickly calculate how long it will take me to drive to Toronto (many assumptions apply of course). Now the 200 km is the size of the work that needs to be accomplished and the 100 km/hr is the rate of productivity. You need BOTH numbers to build a schedule.

In any aspect of software development, be it requirements, design, coding, testing, documentation, indeed the size of the Project Plan, both the size and the rate of productivity can be estimated using Function Points (FP). Normally we measure the rate in Function Points per Month (FPPM). For simplicity, we assume that a "month" of work is 4 five-day weeks or 20 days.

The details of how we calculate FPs is not our concern here. For your information, the world's smallest program, (printf ("Hello World!");) is about 2 FPs. The world's largest, Microsoft

Windows 8, is about 425,000 FPs. All of the rest of the world's programs lie somewhere in between. Most programs are under 5,000 FPs. The production rates are highly variable but here are the rates that we will use. We use them in the sense that if a module is 1000 FPs in size and the rate to perform a procedure is 100 FPPM then it will take 10 person-months to do the thing. Or 10 people can do it in one month if that is possible. Some sample FPPM ratings for the various subcomponents are given next [1].

- Requirements 200 FPPM
- Design 50 FPPM
- Coding 15 FPPM
- Testing 100 FPPM
- Integration 500 FPPM
- Deployment 750 FPPM

These would be used in estimation in the following way. Suppose we were doing requirements for a project that was estimated to be 1000 function points in size. We can execute requirements at the rate of 200 FPPM. Thus it would take us 1000/200 or 5 person-months to get the work done. This is an example of parametric estimation.

6.2.8. Project and resource calendars

Having calculated the duration as shown in the above section, the actual start and finish dates of an activity depend on the layout of the rest of the activities, as well as the constraints of the project and resource calendars. The project calendar prescribes the working/non-working dates and time for the whole project. For example, no team member is going to work on New Year's Day (Jan 1). The team may decide to work overtime for one particular weekend on Saturday, making that day a work day with special work hours accommodating, for example, a database backup at 6 a.m.

Resource calendars, on the other hand, are specific to individual resource(s). For example, a resource who will be on vacation for the next three weeks will have her unfinished activities pushed out by three weeks for completion. Note that activities assigned to other resources with dependencies on these pushed-out activities will be impacted too. Hence, it is a good idea to set up the project and resource calendars first in any schedule. It is then up to the PM to minimize the impact of non-working dates and time on the project.

6.3. Time Executing

In this phase, all schedule activities are being executed in sequence from start to end according to the order given in the AOA/AON diagram. In particular, the PM must be constantly aware of the available buffers and critical paths (paths with no buffer) in order to avoid schedule slippage. A schedule baseline should be taken prior to project execution and status of the schedule is regularly compared with the baseline in order to identify problems and take corrective actions.

To identify available buffers and critical paths in a schedule, the Early Start (ES)/Finish (EF) and Latest Start (LS)/Finish (LF) of an activity (see Figure 6.4), and the forward and backward paths need to be understood and examined in detail (see Figure 6.5(a)).

6.3.1. Forward path calculation

The forward path calculation is based upon the Early Start (ES) and Early Finish (EF) calculation from Start to End. Assume the

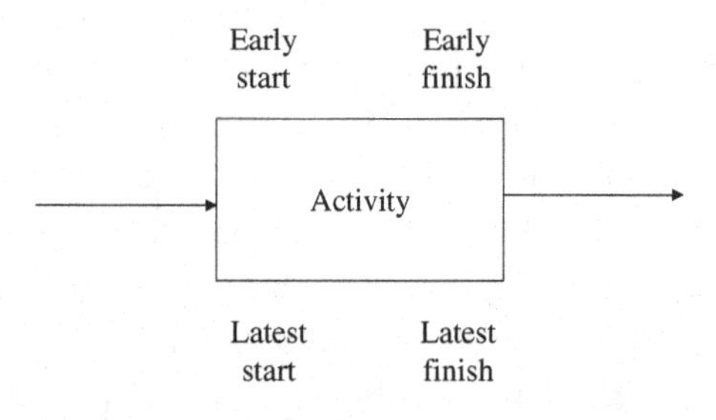

Figure 6.4. ES, EF, LS, LF of an activity.

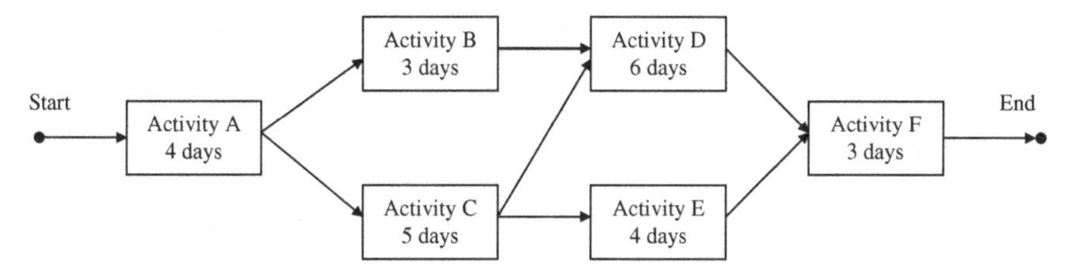

Figure 6.5(a). Forward and backward path example.

schedule starts on day 0, ESs of all activities connected to the start are 0. Within the same activity,

$$EF = ES + \text{duration}$$

ES of subsequent activity is the largest of all EFs of its predecessors because the subsequent activity cannot begin until all preceding schedule activities (predecessors) have been completed. For example, ES of Activity D will take the larger of the EFs of Activities B or C.

Figure 6.5(b) shows the entire length of the project to be 18 days.

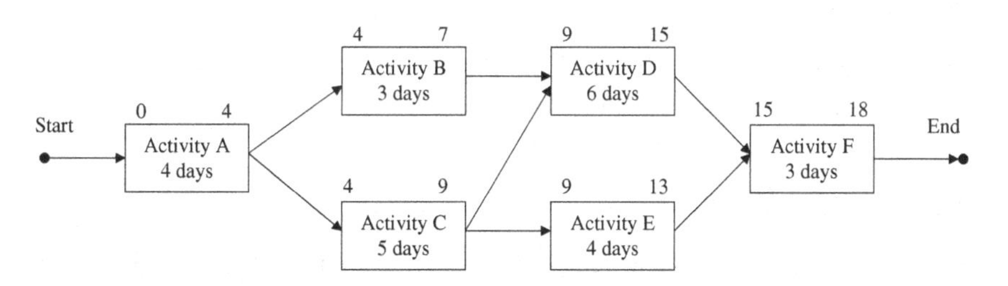

Figure 6.5(b). Forward path calculation.

6.3.2. Backward path calculation

The backward path calculation is based upon the Latest Finish (LF) to Latest Start (LS) calculation from End to Start. Assume the schedule ends on Day D. LFs of all activities connected to the end are D. Within the same activity,

$$LS = LF - \text{duration}$$

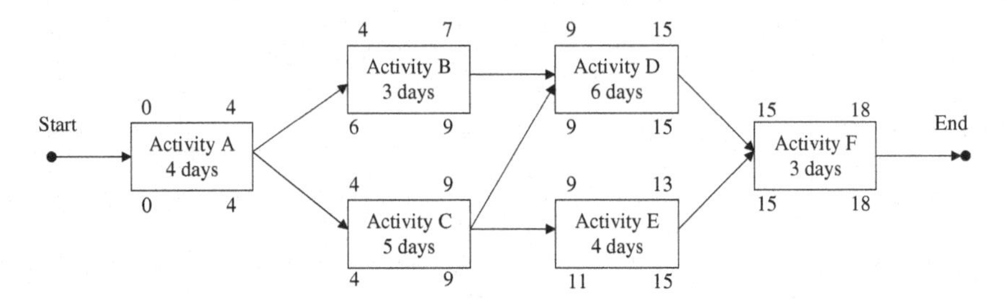

Figure 6.5(c). Backward path calculation.

LF of the preceding activity is the smallest of all LSs of its successors because the activity must end to enable the earliest succeeding schedule activities (successors). For example, LF of Activity C will take the smaller of the LSs of Activities D or E (see Figure 6.5(c)).

The backward path calculation can be used to confirm the correctness of the forward path calculation. If both calculations are correct, within all the starting and ending activities, i.e. Activity A and Activity F in Figure 6.5(c), ES = LS and EF = LF.

The backward path calculation can also be used to evaluate whether an imposed deadline on a project is feasible. For example, can the project represented by Figure 6.5(c) be completed in 15 days? As an exercise, you can simply do a backward path calculation starting from Activity F with LF = 15. You will find that the ES of Activity A is negative, implying mission impossible.

6.3.3. Float (Slack)

Float (or slack) is the amount of time that an activity can be delayed before causing delay on the downstream activities or the entire project (see Figure 6.6(a)).

$$\text{Free Float} = ES_{\text{Activity2}} - EF_{\text{Activity1}}$$

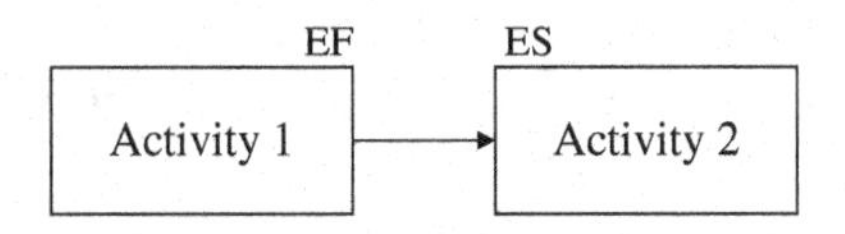

Figure 6.6(a). Free float.

Free float is the buffer between two activities. Suppose there is a free float of D days between Activity1 and Activity 2, Activity 1 has a buffer of D days without impacting Activity 2.

Total Float = LF – EF

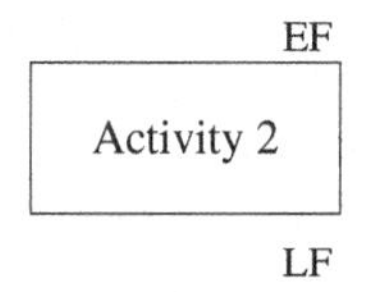

Figure 6.6(b). Total float.

Total float (see Figure 6.6(b)) is the buffer within an activity without delaying the whole project. Continuing with the same example,

Activity B has a free float of 2 days to Activity D

Activity E has a free float of 2 days to Activity F

Activity B has a total float of 2 days

Activity E has a total float of 2 days

Note that it is just coincidence (see Figure 6.6(c)) that free and total floats are equal in this case. There will be an example later to illustrate the difference in free and total floats.

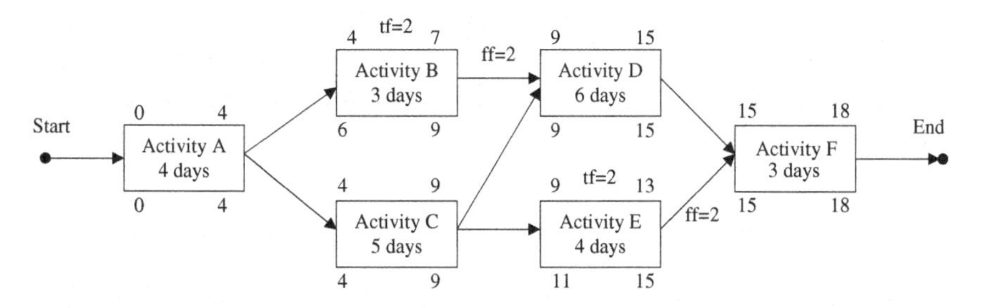

Figure 6.6(c). Free float and total float example.

6.3.4. Critical path

Critical path is the path through the schedule activities with no buffer and it should be carefully monitored, as one day of delay along the critical path means one day of delay on the whole project.

There are two ways to identify the critical path,

- path with 0 float (i.e., no free float and no total float)
- longest duration through the project schedule

The path with zero float is A-C-D-F (see underlined activities in Figure 6.7).

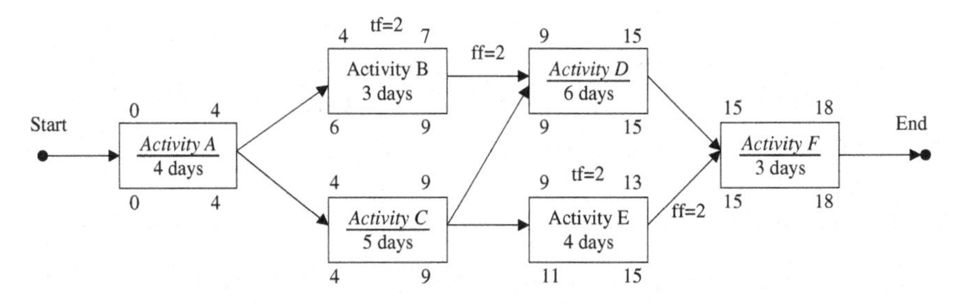

Figure 6.7. Critical path example.

There are three paths through the schedule:

- A-B-D-F 16 days
- A-C-D-F 18 days
- A-C-E-F 16 days

Note that there can be more than one critical path in a schedule, and all critical paths need to be closely monitored. Note also that delaying any activity on a non-critical path long enough may make that path critical.

6.3.5. Lag and lead

A lag inserts a delay in the successor activity and a lead allows an acceleration of the successor activity. (A simple example would be: delay 3 days for the concrete to cure before putting up the frame.) There are four such relationships.

- Start – Start (Figure 6.8(a))
- Finish – Start (Figure 6.8(b))
- Start – Finish (Figure 6.8(c))
- Finish – Finish (Figure 6.8(d))

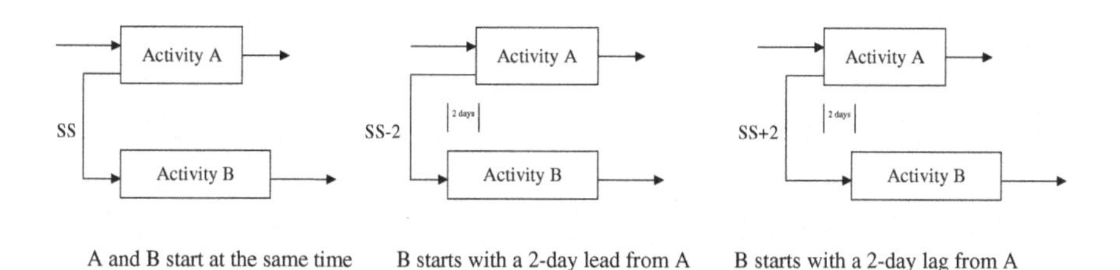

Figure 6.8(a). Start – start.

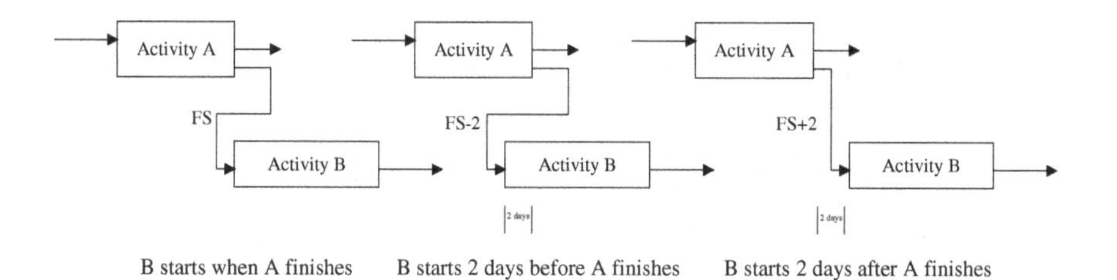

Figure 6.8(b). Finish – start.

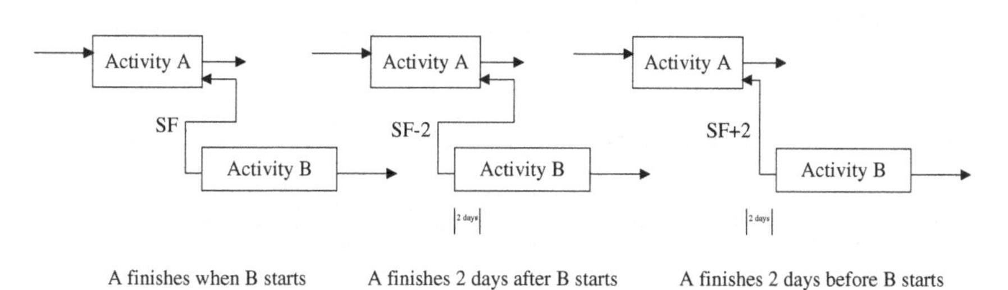

Figure 6.8(c). Start – finish.

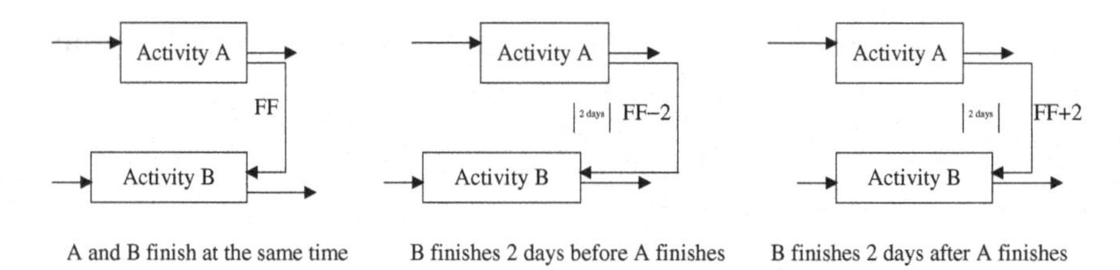

Figure 6.8(d). Finish – finish.

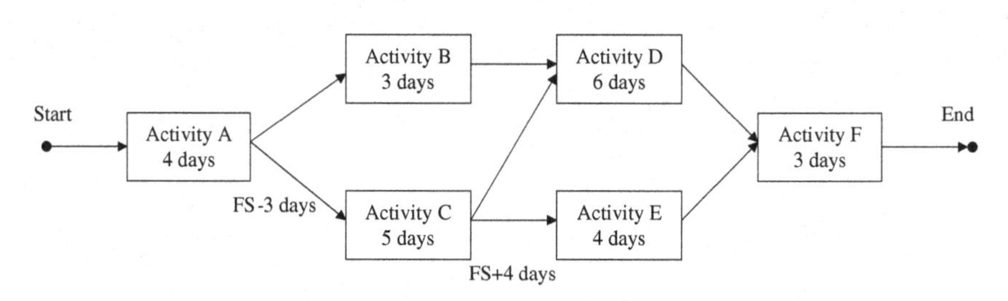

Figure 6.9. Lag and lead example.

Continuing with the previous example, lags and leads are added to the schedule (see Figure 6.9). As an exercise, calculate the forward and backward paths, free and total floats, and critical paths based on the concepts discussed in the previous sections. Compare your results with the following discussions.

Forward Path

$$ES_{\text{Activity C}} = EF_{\text{Activity A}} - 3 \text{ days} = 1$$

$$ES_{\text{Activity E}} = EF_{\text{Activity C}} + 4 \text{ days} = 10$$

Backward Path

$$LF_{\text{Activity C}} = \text{earliest of } (LS_{\text{Activity E}} - 4 \text{ days}, LS_{\text{Activity D}}) = 6$$

$$LF_{\text{Activity A}} = \text{earliest of } (LS_{\text{Activity C}} + 3 \text{ days}, LS_{\text{Activity B}}) = 4$$

Free Float

Activity D has a free float of 1 day to Activity F.

Activity C has a free float of 1 day to Activity D, but no free float (due to FS+4) to Activity E, hence no free float overall.

Total Float

Activity B has total float of 1 day, meaning availability of a 1-day buffer without delaying the whole project. If Activity B finishes 1 day later, $ES_{\text{Activity D}}$ becomes 8 and $EF_{\text{Activity D}}$ becomes 14, $ES_{\text{Activity F}}$ is still 14, which means no delay in the project.

Activity D has total float of 1 day.

Critical Path

The path with zero float is A-C-E-F.

There are three paths through the schedule:

- A-B-D-F 16 days
- A-C-D-F 15 days, including FS-3 from A to C in the calculation
- A-C-E-F 17 days, including FS-3 from A to C, and FS+4 from C to E in the calculation.

6.4. Time Monitoring and Controlling

As mentioned previously, a schedule baseline is developed prior to project execution. The baseline is subject to approved changes to the project scope. The PM monitors the progress of the project and updates the schedule accordingly. The floats and critical paths are monitored and compared with the baseline. The schedule performance index (SPI) is the key indicator of schedule slippage for engaging corrective actions. Earned value and SPI will be covered as methods to confirm progress of the schedule in Chapter 7.

6.4.1. 50-50 Rule

An interesting phenomenon in reporting progress by team members is the percent completion of schedule activities. For example, it is not unusual for a team member to report 50% completion on an activity in one week, then 80% completion the week after (while the work is supposed to be done), 90% completion another week later, followed by 95%, yet the activity never seems to complete. There is the 50-50 rule, or other variations of it, to avoid such problem by forcing work to be reported as 0%, 50%, or 100% complete (i.e., no percentage in between). If a schedule activity is not completed week after week, the slippage is going to surface, regardless of the actual percent completion.

6.4.2. Gantt chart

A Gantt Chart (see Figure 6.10) is a schedule communication tool that provides a snapshot of the progress of schedule activities with completed activities or percentage completion marked as solid bars. Note that the interdependencies amongst activities are not shown (unlike the AON diagram).

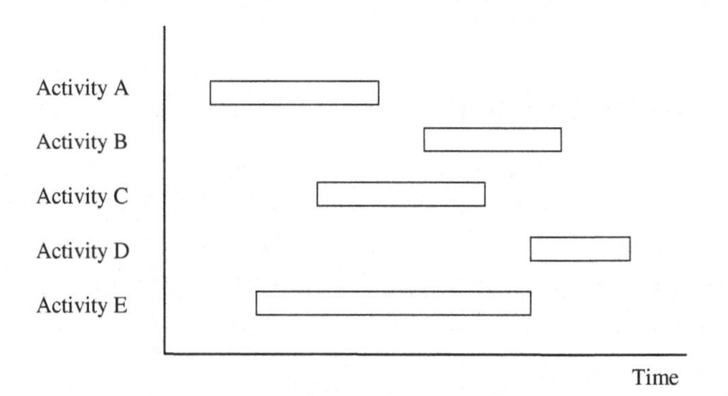

Figure 6.10. Gantt chart example.

6.4.3. Milestone chart

A Milestone Chart (see Figure 6.11) is another reporting tool that provides a snapshot of major accomplishments on planned versus actual dates.

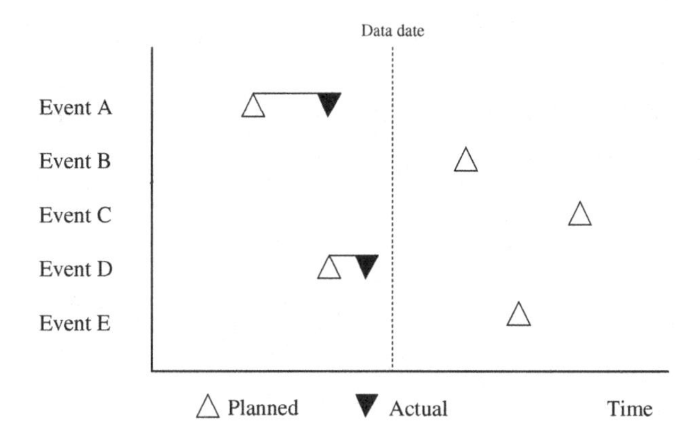

Figure 6.11. Milestone chart example.

6.4.4. Schedule control techniques

Two common techniques to compress schedule and fight slippage are:

- Crashing which means the addition of resources even though productivity will be reduced.
- Fast tracking which means carrying out activities in parallel (e.g., start coding a few days prior to completion of design) even though the risk is increased and rework may result.

6.5. Time Closing

Provide final schedule updates by closing schedule activities as appropriate. The final SPI value (see Chapter 7) and the metrics for schedule performance will be recorded in the Project Plan and in the Lessons Learned document.

6.6. Scheduling Exercise

1. Continuing with the WBS Exercise at the end of Chapter 5, use the work packages derived in the WBS to develop a schedule

with external dependencies and milestones, and with less than 30 activities.

2. Read about Critical Chain as another Schedule Control Technique and give an example.

6.7. Multiple Choice Questions

1. What is the critical path of the following diagram?
   (a) A-B-D-F
   (b) A-B-E-F
   (c) A-C-E-F
   (d) There are more than one critical paths

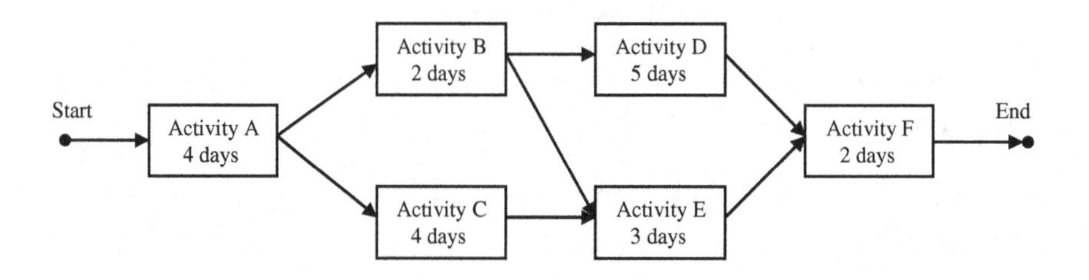

2. You estimated an activity with an effort of 10 staff-days. How many days will it take to complete the activity if you were given 2 resources to work on it, and each resource has a productivity factor of 80%, and availability of only 50%?
   (a) 8 days
   (b) 12.5 days
   (c) 16 days
   (d) 25 days

3. Which project management process does Develop Schedule under Project Time Management belong to?
   (a) Initiation
   (b) Planning
   (c) Execution
   (d) Control

4. For an activity with ES = 4, EF = 7, LS = 6, LF = 9, the activity has:
   (a) Total float of 2
   (b) Free float of 2
   (c) Total float of 3
   (d) Free float of 3

5. Refer to the following diagram:
   (a) A has a free float of 1
   (b) A has a free float of 2
   (c) B has a free float of 3
   (d) C has a total float of 2

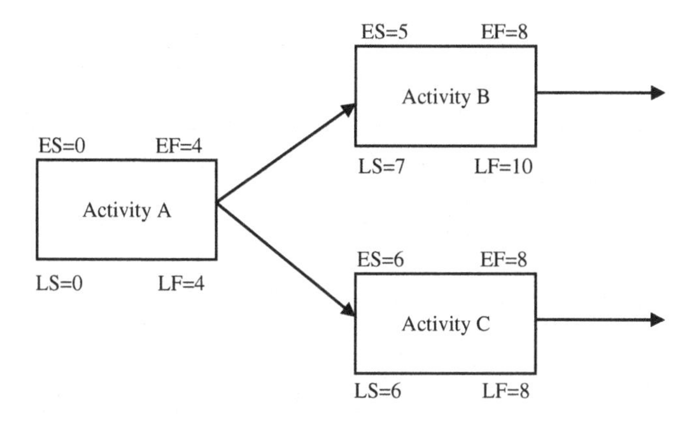

6. A Gantt Chart is a:
   (a) Activity-on-arrow diagram
   (b) Milestone chart
   (c) Schedule communication tool
   (d) Activity-on-node diagram

7. What is the following diagram known as?
   (a) Gantt chart
   (b) Precedence diagram
   (c) Activity-on-arrow diagram
   (d) Activity-on-node diagram

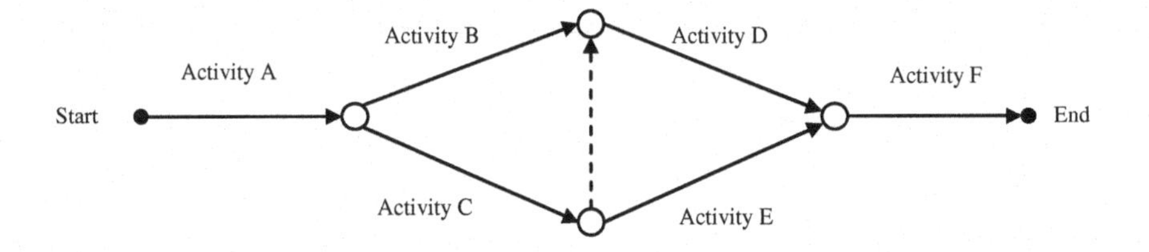

8. Refer to following diagram:
   (a) A has a free float of 2
   (b) A has no free float
   (c) B has a free float of 2
   (d) C has a free float of 2

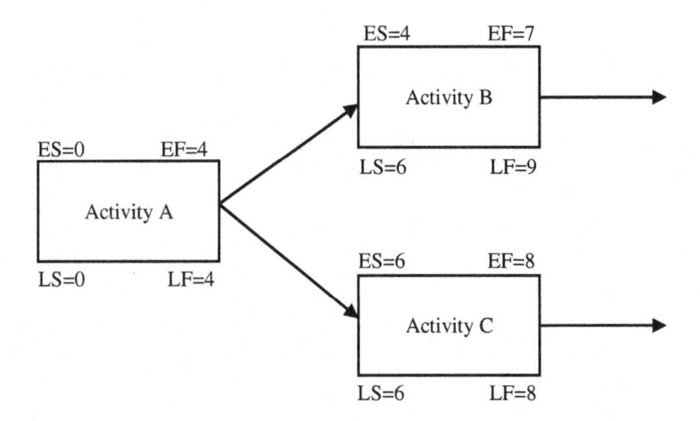

9. Based on an initial planning session, you realize that the project cannot meet the desired end date. You immediately negotiate with the manager of another project to transfer an experienced staff to your project and take over some activities. This technique is known as:
   (a) Crashing
   (b) Fast tracking
   (c) Critical chain
   (d) Optimization

10. A project schedule completion date will change if:
    (a) The critical path is reduced
    (b) The contingency is no longer available
    (c) No float time is available
    (d) Project resources are reduced

11. Your project is behind schedule due to conflict between team members. Having resolved the conflict, to get the project back on schedule, you should consider:
    (a) Crashing the schedule
    (b) Performing resource leveling
    (c) Conducting reverse resource allocation scheduling
    (d) Utilizing the critical chain resources

12. The purpose of fast tracking a project is to:
    (a) Increase productivity
    (b) Reduce project duration
    (c) Increase schedule tracking controls
    (d) Reduce project risks

13. A generally accepted method to confirm accuracy of task progress is through:
    (a) Earned value
    (b) Probability versus outcome
    (c) Maximum ceiling
    (d) Work breakdown structure

14. Which of the following methods is used to control the project schedule:
    (a) Pareto Diagram
    (b) Performance measurement
    (c) Parametric modeling
    (d) Statistical sampling

15. Elements of changing a project schedule include all of the following except:
    (a) Obtain the appropriate levels of approval
    (b) Submit the appropriate change requests
    (c) Evaluate the impact of a change to the schedule
    (d) Adjust the project end date to the schedule variance

References

1. Jones, C. *Estimating Software Costs*, 2nd Edn. McGraw-Hill, 2007.

CHAPTER SEVEN

Cost Management

7.1. The Purpose of Cost Management

Cost management is the twin of time management. In fact, often the activities in Chapters 6 and 7 are combined together into a single activity, particularly in small projects. Cost management includes all the processes necessary to plan, estimate, budget, and control costs. Normally this is one area where organizations excel. If they are in business to make a profit, they are likely to be really good at managing costs. In this chapter, we are primarily concerned with the costing of resources necessary to get the work done, now that we know the amount of work that has to be done. In doing costing, it is important to understand the scope of the costing. Are we concerned just with the cost of the creation of the product, or are we concerned about life cycle costing, sometimes call Total Cost of Ownership? Again, we want to make this clear as to the scope of the cost estimation we are performing.

7.2. Cost Planning

7.2.1. Cost estimating

The first step in cost management planning is to estimate the cost of resources to complete the project. The granularity or precision of the estimate matures over time. Table 7.1 summarizes the stages of cost estimates.

The top-level WBS is used to give an Order-of-Magnitude (OOM) estimate, also known as simple wild ass guess (SWAG). The

Table 7.1. Cost estimates.

Cost Estimates	WBS	Precision
Order-of-Magnitude Estimates	Top	−25%, +75%
Budget Estimates	Middle	−10%, +25%
Definitive Estimates	Low	−5%, +10%

mid-level WBS is used to give a budget estimate, also known as preliminary estimate. The low-level WBS is used to give a definitive or detailed estimate.

7.2.2. Cost estimating methods

Several methods can be used to derive the cost estimates:

- Analogy and Vendor Bid — uses past projects of similar nature to estimate the current project.
- Parametric Modeling — uses mathematical models and formulae to conduct regression analyses of historical data in order to derive a model that predicts the cost estimate of the current and future projects.
- Bottom-up Method — uses detailed scheduling to estimate all schedule activities. The summation of all these will give a cost estimate of the whole project.
- Quantitatively Based — multiply total quantities (e.g., 100 hamburgers) by cost estimate for each (e.g., 3 staff-minutes to make a hamburger) to derive total cost (e.g., $100 \times 3 = 300$ staff-minutes, or 5 staff-hours). This method only applies to linear scale estimates.

$$\text{Total cost} = \text{quantity} \times \text{unit cost}$$

7.2.3. Cost budgeting

There are two types of costs: direct cost (cost directly applicable to the project) and indirect cost (cost not applicable to the project, e.g., executive salaries). Typically, indirect cost is added to the budget as a percentage of the direct cost. Figure 7.1 gives a breakdown of project price.

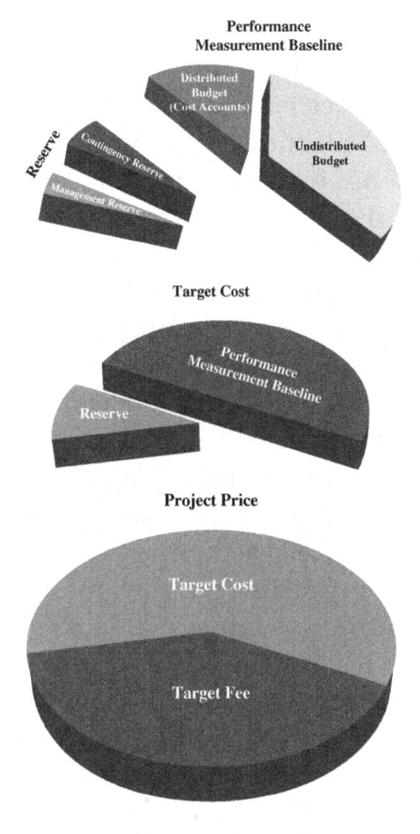

Figure 7.1. Project price.

Contract price is the sum of the target cost (which is the direct cost) incurred by an organization and target fee which is the desired profit. The target cost consists of the performance measurement baseline, which is the cost estimate of the entire project given by some method in the previous section, and reserve, which is the buffer for overrun due to unforeseen circumstances. The undistributed budget consists of work packages which are unopened (for future schedule activities), and the distributed budget, also known as cost accounts, consists of opened work packages with schedule activities under execution. For example, the "Class Diagrams" work package in SPENTFUEL becomes part of the distributed budget when activities 1, 2, 3, or 4 as detailed in Section 6.2.1 are under execution.

The reserve is usually some percentage of the target cost to accommodate overrun of existing work and execution of as yet unidentified work. Percentages will vary from a very small amount (2% to 3%) all the way to 30% [1], depending on the area in which we are executing the project. The better known the processes are, the lower the percentage used for reserve calculation. For example, the reserve in the Confederation Bridge was 2%, while a common reserve for software projects is 30%. There are two types of reserves: contingency reserve and management reserve. Contingency reserve is the known unknown, which means the risk is known to the project team, yet the impact is unknown or unrealized. The contingency reserve is a line item in the project budget. Management reserve is the unknown unknown, which means the risk is unknown to the project team and obviously, the impact is also unknown and unrealized. Management reserve is a line in the corporation's budget, not the project's budget. Remember, many years ago, Ford Motor had to recall all Firestone tires mounted on the Ford Explorer because treads separating caused catastrophic accidents in the SUVs [2]. That was an example of an unknown unknown impact on the budget. Ford had no idea that the tires would fail in such an unfortunate way and it certainly could not gauge the financial impact of those costly multibillion-dollar settlements. These funds must be taken from the management reserve.

7.3. Cost Executing

In this phase, the undistributed budget is gradually turned into the distributed budget as all schedule activities are executed in sequence from start to end. The project manager must be constantly aware of the distributed budget versus the actual spending in order to avoid cost overrun. A cost baseline should be taken prior to project execution and the reserves and budget (distributed and undistributed) must be monitored and controlled.

7.4. Cost Monitoring and Controlling

Earned value management is the most important concept in cost (and schedule) control. By now, we have completed our WBS and we have scheduled our activities. We can construct our S-curve and start the project! Suppose we have spent 3 months on the project and have spent $100,000 executing our work. Our S-curve also says we should have spent $100,000 by the end of month 3. Are we on schedule? Are our costs appropriate? Maybe, if we are really lucky. The problem is that we do not have the right information. Suppose our project manager, Bad Bob, flew us to Jamaica for 3 months at a cost of $100,000. No work done!

What is missing is the concept of work well done. How do we check that? At our progress sign-off, we present our deliverables that get signed-off if they are done well. Thus we have three items of interest. The Planned Value (PV) is the work, measured in dollars, we estimated it would take to complete the element in building the details of the WBS. The S-curve shows all of the PV over the timeline of the project. The Actual Cost (AC) is the amount we really spent in completing the work element which can be the same, greater, or less, depending on conditions. Finally, the Earned Value (EV) is the amount we can "earn" by getting sign-off. Note that it is the PM who estimates the PV and hence the EV.

Consider the following terms:

- Planned Value (PV) — budget originally planned for
- Earned Value (EV) — value of work earned from the plan upon completion of certain schedule activities
- Actual Cost (AC) — actual cost incurred during execution of certain activities
- Cost Variance CV = EV – AC
  (0 => on cost, <0 => overrun, >0 => under run)
- Schedule Variance SV = EV – PV
  (0 => on schedule, <0 => behind schedule, >0 => ahead of schedule)

- Cost Performance Index CPI = EV/AC
  (1 => on cost, <1 => overrun, >1 => under run)
- Schedule Performance Index SPI = EV/PV
  (1 => on schedule, <1 => behind schedule, >1 => ahead of schedule)

CPI and SPI are two key indicators in cost and schedule control. Well-accepted industrial standards would tolerate a ±10% (i.e., 0.9–1.1) CPI and SPI, 10–20% will raise a yellow flag, whereas >20% will raise a red flag. Note that too big a deviation, despite ahead of schedule and cost under run, would warrant an examination of the schedule and cost baselines. (Is the schedule too relaxed? Are there any missing schedule activities from the work packages?)

Consider the following example:

As can be seen from the graph (in Figure 7.2(a)), the project is behind schedule and experiencing cost overrun.

Figure 7.2(b) pictorially shows the relationships amongst the various terms for earned value management.

Here is an EV exercise for you to try:

Mr. Hug Atree hired you to plant trees for 10 days. You are supposed to plant 10 trees per day and he pays you $10 for each tree. After 4 days, you have only planted 30 trees, yet you

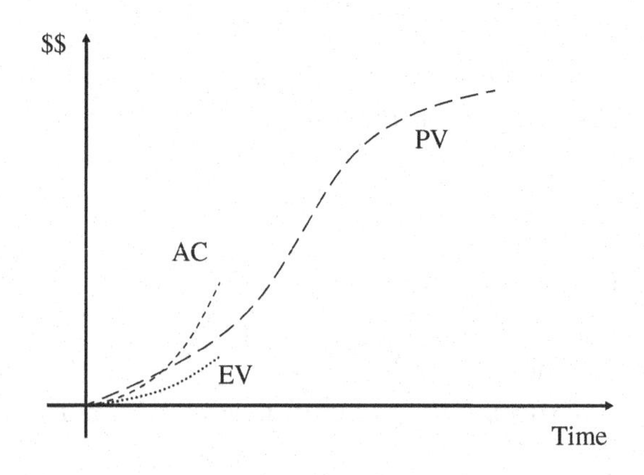

Figure 7.2(a). Earned value example.

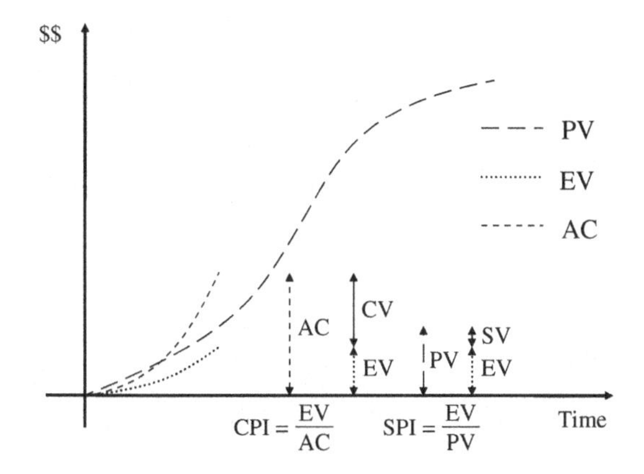

Figure 7.2(b). Earned value management terms.

reported that you have put in a lot of overtime approximating an extra day of work. Mr. Atree was very generous to pay you $500 ($400 for 4 days of work, despite that you have only planted 30 trees, instead of 40; and $100 in compensation of your overtime effort). After 4 days:

- Planned value PV =
- Earned value EV =
- Actual cost AC =
- Cost variance CV = EV – AC =
- Schedule variance SV = EV – PV =
- Cost performance index CPI = EV/AC =
- Schedule performance index SPI = EV/PV =

Continuing with the same exercise, more useful terms are introduced (see Figure 7.2(c)):

- Budget at completion BAC =
- Percent complete = EV/BAC =
- Percent spent = AC/BAC =
- Estimate at completion EAC = BAC/CPI =

Based on the current spending habit, how much does it cost to complete the project?

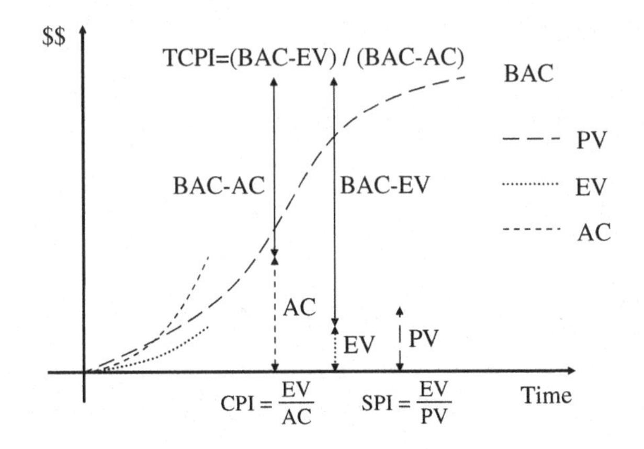

Figure 7.2(c). TCPI relationship.

- Estimate to complete ETC = EAC – AC =

  Based on the current spending habit, how much more does it cost to complete the project?
- To-complete performance index

  TCPI = (BAC – EV)/(BAC – AC) =

  TCPI is somewhat like the cost performance index for the remainder of the project in order to complete within cost. If there is already cost overrun, how much does the belt have to be tightened in order to still finish the project within the original budget? If TCPI results in weird values (e.g., negative, infinity), the original budget cannot be maintained.

If one accepts the fact that BAC cannot be maintained, the project will work towards EAC, then TCPI = (EAC – EV)/(EAC – AC).

7.5. Cost Closing

The final activity in Cost Management is to record the final CPI, both in the Project Plan and in the Lessons Learned document.

7.6. Multiple Choice Questions

1. You have been hired to plant trees for 10 days. You are supposed to plant 10 trees per day and you are paid $10 for each tree planted. After 4 days, you planted 50 trees and you were paid $500. What are the SPI and CPI for the project?
   (a) SPI = 0.8, CPI = 1.0
   (b) SPI = 1.25, CPI = 1.0
   (c) SPI = 1.0, CPI = 0.8
   (d) SPI = 1.0, CPI = 1.25

2. Which of the following is not a direct cost?
   (a) Labour
   (b) Travel
   (c) Material
   (d) Administrative overhead

3. For a project with SPI = 1.60 and CPI = 2.0, what should you as the project manager do?
   (a) Discuss rationale with the team and take corrective actions
   (b) Relax and go for a celebration because you are ahead of schedule and below cost
   (c) Examine the WBS and ensure that there are no missing activities, assign additional responsibilities to individuals, re-baseline the schedule if appropriate, since the original baseline might be too conservative
   (d) Do nothing

4. Which of the following is known unknown?
   (a) Distributed budget
   (b) Undistributed budget
   (c) Management reserve
   (d) Contingency reserve

5. Which of the following applies to opened work packages?
   (a) Distributed budget
   (b) Undistributed budget
   (c) Reserve
   (d) Target fee

6. For a project with SPI = 0.90 and CPI = 0.8, what should you as the project manager do?
   (a) Discuss rationale with the team and take corrective actions
   (b) Thank the team, relax and go for a celebration
   (c) Re-baseline the schedule, since the original baseline might be too aggressive
   (d) Do nothing

7. For a project with SV = –$200 and CV = –$200, what conclusion can be made on the project status?
   (a) You are ahead of schedule and below budget
   (b) You are behind schedule and over budget
   (c) You are on schedule and on budget
   (d) None of the above

8. You estimated an activity with an effort of 10 staff-days. If you were given two resources to work on it, and each resource has a productivity factor of 80%, and each day you pay $200 for each resource, how much do you need to budget for the activity?
   (a) $625
   (b) $1250
   (c) $2500
   (d) $5000

9. Earned value performance data can be compared to all of the following project management tools except:
   (a) Critical path analysis
   (b) Technical performance metrics
   (c) Risk mitigation plans
   (d) Forecasted final costs and schedule estimates

10. The measure used to forecast project cost at completion is:
    (a) CPI
    (b) SPI
    (c) EV
    (d) AC

11. A project was estimated to cost $1.5 million and scheduled to last six months. After three months, EV = $650,000 PV = $750,000 AC = $800,000; what are the schedule and cost variances?
    (a) SV = +$100,000, CV = +150,000
    (b) SV = +$150,000, CV = –$100,000
    (c) SV = –$50,000, CV = +150,000
    (d) SV = –$100,000, CV = –$150,000

12. You have been contracted to manage a project. The estimated cost of the project is $1,000,000. Your earned value calculation indicates that the project will be completed on time and under budget by $200,000. Your personal profit will decrease by $2,000.

    At the completion of this project, you will document and archive all project information. This costing information may be used for future projects in all areas except to:
    (a) Estimate durations
    (b) Administer contracts
    (c) Resolve conflicts
    (d) Allocate resources

13. What is the precision of a preliminary estimate?
    (a) +/– 50%
    (b) –25%/+75%
    (c) –10%/+25%
    (d) –5%/+10%

14. Which of the following is not a cost estimating method?
    (a) Parametric modeling
    (b) Vendor bids
    (c) Bottom-up
    (d) Critical chain

15. You have been contracted to manage a project. The estimated cost of the project is $1,000,000. Your earned value calculation indicates that the project will be completed on time and under budget by $200,000. Your personal profit will decrease by $2,000.
    Given the estimated decrease in personal profit, what action should you take?
    (a) Invoice for the full $1,000,000 based on the contract
    (b) Add tasks to improve the outcome and increase the actual project cost
    (c) Inform the end-user that you can add features to the project in order to use the entire budget
    (d) Communicate the projected financial outcome to the project sponsor

References

1. Private communication from Government of Canada Treasury Board expert, 1996.
2. Available at: http://en.wikipedia.org/wiki/Firestone_and_Ford_tire_controversy. Accessed February 2013.

CHAPTER EIGHT

Quality Management

What is quality? It is the degree to which a set of inherent characteristics fulfill requirements. It is also the totality of characteristics of an entity that bear on its ability to satisfy stated or implied needs. When we talk about quality, there are two dimensions that we are concerned with. The first is the quality of the product that we are building. It will have a set of quality characteristics that we have built into the product to keep the customers happy. The second dimension is the quality of the execution of the project. As we run the project, are we following proper project management procedures? Are we maintaining earned value, for example? We will deal with both of these aspects of quality as we proceed through the following sections.

8.1. Introduction to Quality

In a seminal paper in 1984, David Garvin examined in fine detail the question of quality [1]. Why was quality so hard to define? Garvin answered that question by pointing out that there were five different ways that we could look at quality, depending on what our perspective or point of view was. Those five ways are:

- transcendental
- user view
- manufacturing view
- product view
- value-based view

The first view is the Transcendental View. Quality can be recognized in an abstract sense, but it cannot be precisely defined. It is like Plato's description of the ideal or any religious leader's description of a quality human existence. It is an ideal to which we can strive but which we will never attain. It is an ethereal view, and while it may be very pleasing philosophically, it is of little use in our practice of project management. The second view is the User View. This is the concrete view of the product grounded in the characteristics of the product to meet the users' needs. It evaluates the product in a task context. Is the user happy using the product? The manufacturing method or characteristics of the product are not important as long as the users are pleased. The third view is the Manufacturing View. This focuses on the product quality during and after production. It leads to a quality assurance that is almost independent of the construction of the product itself. ISO9001:1994 (specific standard covering quality management in the design and development areas, including project management) and the Capability Maturity Model (CMM) [2] are two good examples. These manufacturing standards dictate the generic process of building a quality product. What you are producing is irrelevant. These quality standards in manufacturing are consistent for all manufacturing processes. The fourth view is the Product View. It looks at the inside of the product and assumes that if we control internal product properties, better external properties will result. This leads to quality standards such as Six Sigma, for example, which concentrate on getting rid of all possible defects during the construction of the product itself. The final view is the Value-Based View. Basically, it says if you pay more money for something, it should have better value than a product which costs less. If one buys a Cadillac, one expects much superior performance over an Eastern European car, such as a Yugo, since the Cadillac costs 10 times what the Yugo does. Thus, David Garvin defines five points of view, depending on how one is going to look at the product and how we define quality.

Quality management and project management have the same goal: to produce a product with the appropriate quality using the appropriate processes. Quality is everybody's business, but especially the business of upper management. We can view management's approach to quality as four pillars supporting the quality initiative. The first pillar is **customer satisfaction**. We want to build a product that the customer is excited about. Ed Yourdon coined the phrase "good enough quality". It is important that we understand what the appropriate level of quality is needed. We must define precisely, and with metrics, the level of quality needed in both the product and the project. Take the example of creating a software program. It has been said that there has never been a piece of software written that is error-free. That means we could work on the piece of software literally forever, trying to get that last error out. In the meantime, our competitors in the marketplace would swiftly steal our customers and render us and our organization bankrupt. The second pillar is **prevention over inspection**. Again, to quote a maxim from software engineering: you cannot test quality into bad software. Effort spent in quality prevention will lead to an overall decrease in rework and nonconformance costs. Quality design is absolutely imperative if we desire quality products. The third pillar is the concept of **continuous improvement**, which comes directly from the quality schools. When we discover a defect, it means that we have made a mistake in one of our processes. We need to reverse-engineer our process execution and trace the defect back to its initial creation, determine its root cause, and change the offending procedure to make sure that that particular class of defect can never happen again. We are constantly checking for defects and constantly improving our processes, realizing that although we can never, ever achieve perfection, we can constantly improve what we are doing. The fourth and final pillar of our edifice is **management responsibility**. It is vital that quality be controlled from the highest level of the organization. Everyone, especially upper management, must support the pursuit of quality and champion the quality initiatives, if they are to be successful. If these four fundamental quality pillars are in place, we indeed will produce a quality product by a quality project team.

Now we proceed to try to define quality metrics and quantify what we mean by quality. That will lead into our construction of the quality management plan.

8.2. Quality Planning

In this phase, the relevant quality standards that we will use for our quality work will be identified. Quality standards come from many sources and will be mandated by senior management, government regulations, and end-user requirements. Some examples of quality standards that we could use include ISO 9000, ISO 10006, Malcolm Baldridge, SEI-CMM [2], Deming [3], European Quality Award, Total Quality Management, Six Sigma, Juran [4], Crosby [5], Failure Mode and Effect Analysis (FMEA), Organizational Project Management Maturity Model (OPM3), etc.

Requirements of both the project and the product serve as the baseline to develop quality standards, measurement, and performance. Requirements will originate from stakeholders such as senior management and customers. Stakeholders can be placed into the following categories:

- External — the outside entity that directly pays for the product or service.
- Internal — the inside entity that directly pays for the product or service.
- Intermediate — the entity that uses or tests the product or service prior to final delivery to the external or internal customer (e.g., system test group).
- End user — the entity that does not pay for the product or service, but ends up using it regularly. Note that although this group does not pay, it has a strong influence on acceptance and satisfaction of the product or service.
- Invisible — the entity that the project team is not even aware of (e.g., the environmentalists who try to influence structural engineers in constructing a dam).

8.2.1. Quality metrics

How can we measure quality? Here is a specific example taken from software engineering [6]: the McCall Software Quality metric (aka McCall Index). In a seminal paper, McCall defined three fundamental characteristics that a software product should have from a functional execution point of view. They are:

- product revision
- product transition
- product operations

For each of these three fundamental characteristics, he defined sub-characteristics or as he termed them factors, that each should have.

Product revision talks about the ability to change the software product when errors are found. The three factors that he defined for product revision were:

1. maintenance: how easy is it to fix?
2. flexibility: how easy is it to change it?
3. testability: how easy is it to test it?

Similarly, the three factors he defined for product transition were:

4. interoperability: can it be interfaced to another system easily?
5. portability: can it be transferred to another environment or machine?
6. reusability: can some of the software be reused?

For the final characteristic, product operations, he defined five factors. They were:

7. correctness: does it do what we wanted it to do?
8. reliability: does it do it accurately?
9. efficiency: does it not waste resources?

10. integrity: is it secure?
11. usability: is it user-friendly?

Then he defined metrics for each of the 11 factors. Each metric had to be assigned a value from 0 to 10, where 0 meant that the factor was dreadful, and 10 meant that the factor was superb. Then for each factor, he defined a weighting factor c_j such that the sum of the 11 c_j's were normalized or summed to equal to one. This then resulted in a quality weighting that could range from 0 for a product with abysmal quality to 10 for a perfect product. One would select the weighting factors to show the relative importance of each one of those factors. For example, suppose reusability, correctness, and reliability were the only important factors for a particular software product. Then all of the c_j's would be set to zero, except for c_6, c_7 and c_8, which would all be set to one-third.

The point of discussing McCall's metric is that it can be adapted to any project or product for the measurement of its quality. It also enables us to compare different projects with different weighting factors. The assignation of the metric of 0 to 10 is quite complicated and we refer the reader to McCall's original paper to see the details.

Based on the work of McCall, ISO decided to define its own version of a generic quality metric called ISO 9126 [7]. In this particular standard, ISO defined six primary quality attributes as follows:

1. usability
2. functionality
3. reliability
4. efficiency
5. maintainability
6. portability

Each attribute was then sub-divided into sub-attributes. Usability-defined sub-attributes are those which bear in the effort needed for customer usage and are scored on the individual assessment of such use by those customers. These sub-attributes include:

- understandability
- learnability
- operability

Functionality is defined to be the sum of sub-attributes which bear on the existence of a set of functions and their specific properties. Sub-attributes include:

- suitability
- accuracy
- interoperability
- security

Reliability is defined as attributes that bear on the capability of software to maintain its performance level under stated conditions for a stated period of time. The sub-attributes of reliability include:

- maturity
- fault tolerance
- recoverability

Efficiency are attributes that bear on the relationship between the software's performance in the amount of resources used under stated conditions. Its sub-attributes include:

- time behaviour
- resource behaviour

Maintainability are attributes that bear on making modifications to the software. Its sub-attributes include:

- analyzability
- changeability

- stability
- testability

Finally, portability has attributes that bear on the ability of the software to be transferred from one location to another. Sub-attributes include:

- adaptability
- installability
- conformance
- replaceability

All of these attributes and sub-attributes would then have weighting factors applied, exactly as McCall weighting factors were. It is an alternate way of describing quality and has found favour in many software organizations. Note that the idea behind both McCall and ISO 9126 can be used to define any set of attributes associated with the quality of any engineering product, not just software products.

Most of the metrics we will be defining and using are self-explanatory. We will use metrics such as defects per day or defects per month, for example, to measure our lack of quality. Sometimes we have to measure something that is not quantitative. For example, suppose we wanted to measure the user-friendliness of a piece of software. How would we do that? In order to measure attributes that are on the surface, not quantitative, we need to make them quantitative and we use the Likert scale as defined in Chapter 4.

8.2.2. Quality planning tools and techniques

Several tools and techniques can be used in quality planning. The list given below is by no means complete.

- Benefit/cost analysis — look at the cost required to achieve a certain quality goal and the pay back (benefit). For example, should an organization go up the SEI-CMM ladder (see Appendix D) and set this as a goal? What kind of productivity and quality improvements are going to result?

- Benchmarking

Two or more organizations (even competitors) agree to compare certain aspects (e.g., quality goals, market capture) with each other with the goal of adopting best practices of the other organizations for continuous improvement. Typically, the organizations will sign non-disclosure agreements first and engage a third party to conduct the benchmarking.

- Design of Experiment (DoE)

These are statistical models that help identify which variables have the most influence on the overall outcome.

- Cost of Quality (CoQ)

This is the cost incurred in the budget for assuring or improving conformance to (i.e., keeping errors out of) the process of the quality standards, and for taking corrective actions in fixing non-conformance problems.

  - Conformance
    - Prevention — cost incurred in improvement initiatives to prevent problems from occurring, for example, move up the SEI- CMM ladder.
    - Appraisal — cost incurred in verifying the artifacts produced for the project, for example, formal review of design documents, testing the system prior to delivery to the customer.
  - Non-conformance
    - Internal Failure — cost incurred to fix problems (including re-design, re-implement, re-test) discovered prior to delivery of the product or service to the customer.
    - External Failure — cost incurred to fix problems discovered by the customer.

Note that the non-conformance cost of quality should be kept as low as possible. A typical goal will be around 10–15% of the total

Table 8.1. Cost of quality examples.

Cost of Conformance	Cost of Non-conformance
Money spent during the project, to avoid failure	*Money spent during and after the project because of failures*
Prevention costs (Build a quality product)	Internal failure costs (Failures found by the project)
• training • document processes	• rework
• equipment	• scrap
• time to do it right	
Appraisal costs (Assess the quality)	External failure costs (Failures found by the customer)
• testing	• liabilities
• destructive testing loss	• warranty work
• inspections	• lost business

project budget. The projected conformance cost of quality is typically set at 20–30%, with prevention set at 5–10%.

Table 8.1 provides some illustrative examples [8].

- Control Charts

A control chart is a quality tool designed to display the natural variation or lack of variation in a quality metric. A typical quality metric might be the number of defects per thousand lines of code measured daily. Even if we have a well-defined process to develop code, there will be natural variations from day-to-day in the creation of that product. Some days, the team will do very well; other days, not so well. The idea behind a control chart is to map out the data over a period of time and see if our process is in control or not. We do that by taking our measurements over a period of time, typically at least six time periods, and plotting them on a horizontal timeline as in Figure 8.1. So in our SPENTFUEL example, we will have measured the number of defects per thousand lines of code for several days. We then plot these on a graph. The center line

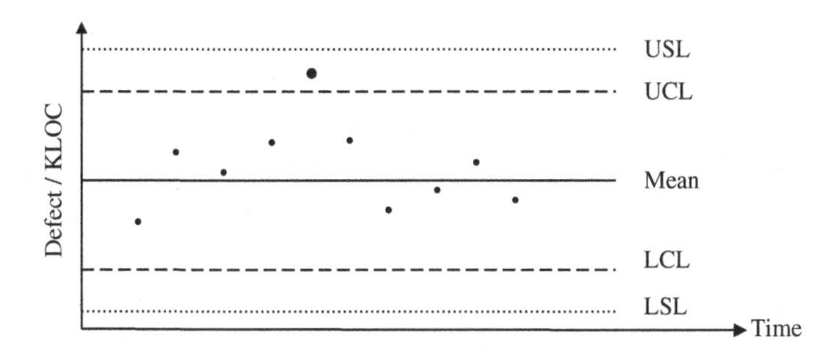

Figure 8.1. SPENTFUEL control chart example.

(CL or mean) is the average of all our measurements. Then we plot the upper control line (UCL) and the lower control line (LCL), which are set at plus or minus 3 Sigma from the mean. Why three Sigma? Three Sigma corresponds to a quality approach of approximately 1 defect per thousand opportunities. In engineering, that is a standard design goal for many products. If you take a look at the closest door to you right now, you will see that that door is probably about 2 m tall. It is designed to be three Sigma in terms of allowing people into the room without bowing their heads. Three Sigma means that for every 1000 opportunities, only one is outside of three Sigma. In other words, 999 people can walk through that door and not bang their heads. One poor fellow in 1000 will have to bow his head or get a cranial goose egg. Three Sigma was defined by Shewhart [2], a quality grandfather, in the early 1920s, and it has been commonly used in control charts since then. Two other outside bands are also plotted, the upper specification limit (USL) and the lower specification limit (LSL). These are the conformance specifications for the product itself, as defined by the customer, which may be greater or less than three Sigma. To commence with the construction of the control chart example, we measure the number of defects over a period of time. If all of the defects are within three Sigma of the mean, we state that the process is in control. That means the random variation that occurs in our process is acceptable. If one of the points lies beyond three Sigma or beyond minus three Sigma, we state that the process is out of control. Now we have to go and find the root cause as to what is putting the process out of control. Control charts are very useful tools for displaying the stability of our processes. As we can see from the SPENTFUEL control chart in Figure 8.1, the fifth point is out of the UCL, and therefore our process is out of control. (The definition of out of control is a little more complex.)

Typical causes of variation are common or random causes (business as usual, somewhat expected) and unassigned causes (exceptions). Any deviation above or below the UCL or LCL, respectively, warrants an investigation.. If seven or more consecutive observations are on one side of the mean, despite within the control limits, we also say the process is out of control. Another good example of the use of control charts is to plot SPI and CPI over time. The idea is to take corrective actions before the USL and LSL limits are breached.

8.2.3. Quality planning output

There are five major outputs from quality planning:

1. The Quality Management Plan (QMP)
2. Quality metrics
3. Quality checklists
4. Process improvement plan
5. Project document updates

8.2.3.1. *The quality management plan*

The main output from the planning phase is the quality management plan. Here is a suggested outline for constructing the plan [9].

1. QA Organization and Resources
   - Organization structure
   - Personnel skill level and qualification
   - Resources

2. QA standards, procedures, policies, and guidelines
3. QA documentation requirements
4. QA software requirements
5. Evaluation of storage, handling, and delivery
6. Reviews and audits
7. Configuration requirements
8. Problem reporting and corrective action
9. Evaluation of test procedures
10. Tools, techniques, and technologies
11. Quality control of contractors, vendors, suppliers
12. Additional control
    - Miscellaneous control procedures
    - Project specific control
13. QA reporting, records, and documentation
    - Status reporting procedures
    - Maintenance
    - Storage and security
14. Retention period

8.2.3.2. *Quality metrics*

Quality must be defined in a quantitative way. Some particular attributes that relate to quantity need to be described as metrics or numeric values. We need to define these for the quality plan and build them into the plan itself as to how we are going to control them. For example, the SPI should be measured regularly and the quality plan is to keep that within 10%. Other examples of project quality metrics include:

- schedule control
- cost control
- failure rate, Mean-Time-To-Failure (MTTF)
- defect frequency
- reliability
- test coverage, etc.

8.2.3.3. *Quality checklist*

According to the PMBOK, a checklist is a structured tool, usually component-specific, used to verify that a set of required steps is being performed. Here is an example:

Quality Checklist for Requirements

1. Is analysis of the information domain complete, consistent and accurate?
2. Is problem partitioning complete?
3. Are external and internal interfaces properly defined?
4. Does the data model properly reflect data objects, their attributes, relationships?
5. Are all the requirements traceable to system level?
6. Has prototyping been conducted for the user?
7. Is performance achievable within the constraints imposed by other system elements?
8. Are requirements consistent with schedule, resources, budget?
9. Are validation criteria complete?

8.2.3.4. *Process improvement plan*

The process improvement plan is a sub-plan of the project management plan (see Chapter 3). It details the steps for analyzing and improving processes to enhance project value. Areas we can consider include:

- process boundaries
- process configuration
- process metrics
- targets for improved performance

8.2.3.5. *Project document updates*

Our quality work may necessitate changing other project documents, such as:

- stakeholder registry
- responsibility assignment matrix

- risk registry
- work breakdown structure, etc.

8.3. Quality Executing

In this phase, quality assurance is performed to apply the planned and systematic quality activities to guarantee that the project employs appropriate quality standards and metrics to satisfy quality requirements. Quality audits are conducted to identify inefficient and ineffective policies, processes and procedures in use on the project. Quality audits can be conducted in the following modalities:

- regular or episodically
- by an in-house auditor or by an external agency

8.3.1. The quality audit

The quality audit itself should have accepted guidelines for execution. Just like any other process, quality auditing is mandatory for ISO 9000 [10]. To check out your quality processes practiced, ask the following questions:

1. What are the policies, standards, procedures?
2. Are they written down, followed, checked?
3. Where does QA reside in the organization chart?
4. How are the QA functional interfaces?
5. Does QA interact with formal reviews, configuration management, testing?

The benefits of doing quality audits include:

- reduced software defects
- lower maintenance costs
- improved reputation
- higher reliability (save on legal costs!)

- drop in overall life cycle costs
- increase in people's pride
- identification of good practices being implemented
- identification of gaps/shortcomings
- addition of practical information to the lessons learned repository

However, there are some disadvantages to audits:

- requires upfront dollars (hard to justify)
- hard in small institutions
- need one quality engineer per one hundred regular engineers approximately
- represents a cultural *change* (change is never easy)

In summary, overall project quality performance needs to be evaluated on a regular basis.

8.4. Quality Monitoring and Controlling

In this phase, specific project results are monitored to determine if they comply with relevant quality standards and ways to estimate the cause of unsatisfactory results will be identified. The following list of tools and techniques (by no means complete) can be used in quality control.

- Flow Chart
- Sampling
  - Attribute sampling refers to whether a sample satisfies the requirement (yes or no).
  - Variable sampling refers to specific measurements on a sample (e.g., by weighing it).
- Reviews, Inspections and Walk-throughs (see Section 8.4.1)
- Control Charts (see Section 8.2.2)
- Histograms (a bar chart that measures the frequency of occurrences)
- Pareto Diagram

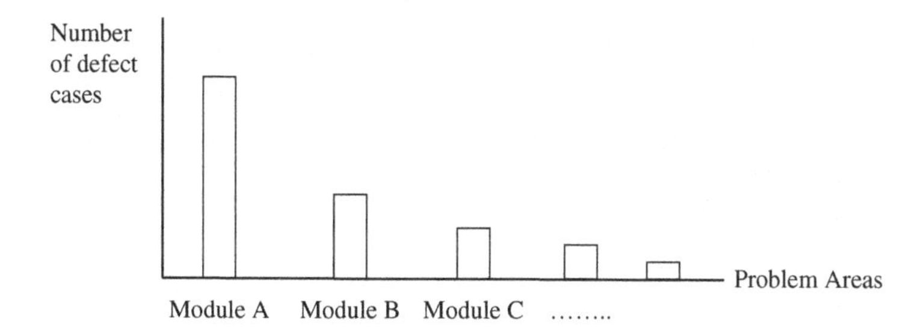

Figure 8.2. SPENTFUEL Pareto diagram example.

This is a histogram ordered in decreasing frequency of occurrence. Pareto was an early 20th century Italian economist. He was the one who first pointed out that in analyzing defects, the relative importance of defects to the project were not uniformly distributed. In fact, he invented what we call the 80–20 rule. The 80–20 rule says that 80% of the trouble is caused by 20% of the causes. Another way of phrasing the Pareto rule is, we want to be concerned about the important few not the trivial many when it comes to being concerned about classes of defects. For example, look at Figure 8.2 that plots the number of defect cases for various modules within the SPENTFUEL Control Module. In it, you will see that Module A has by far the greatest number of defects. Where do you think one should look now to find the next defect? Module A of course! Software defects are highly correlated. The process that enabled you to create the defects that you have discovered probably caused you to create many more that you just have not found yet.

- Cause and Effect Diagram (see example in Figure 8.3)

Also known as the Ishikawa Diagram [11,12] or Fish Bone Diagram, the convention is to write down the effect at the fish head. Then, think about what is causing the effect and keep asking "why" or "how" until the root cause is discovered. One organization that we know of suggested the use of 6 'M's: machine, material, manpower, method, measurement, and motherhood (environment) as

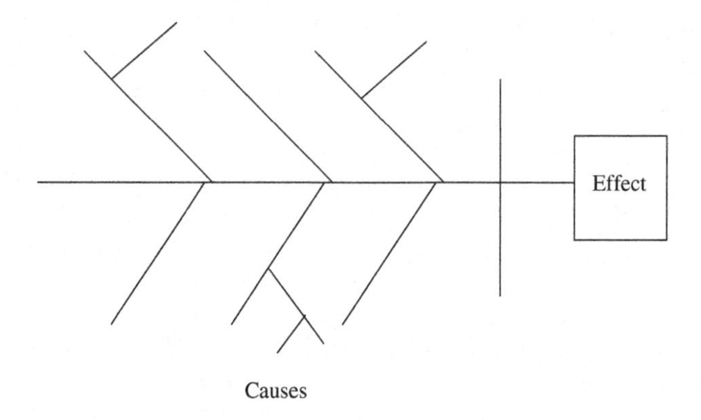

Figure 8.3. Cause and effect diagram.

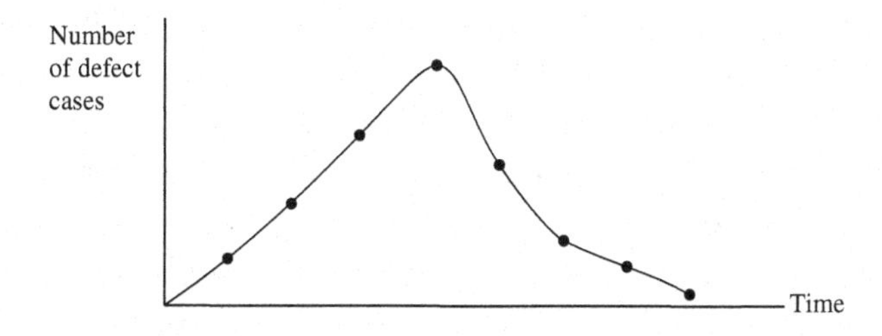

Figure 8.4. SPENTFUEL run chart example.

effects on the fish bones. Next, try the SPENTFUEL Cause and Effect Diagram exercise in Section 8.6.

- Run Chart (see example in Figure 8.4)

This is a graphical display of history and pattern of variance over time. Based on that, trend analyses can be performed.

- Check Sheet (see example in Figure 8.5)

This is a simple counting of occurrences.

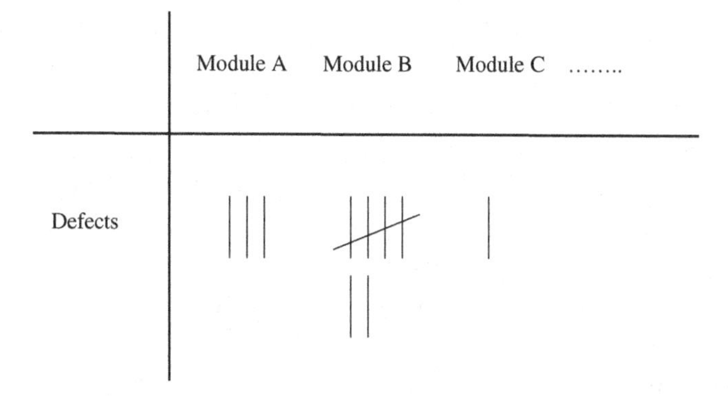

Figure 8.5. SPENTFUEL check sheet example.

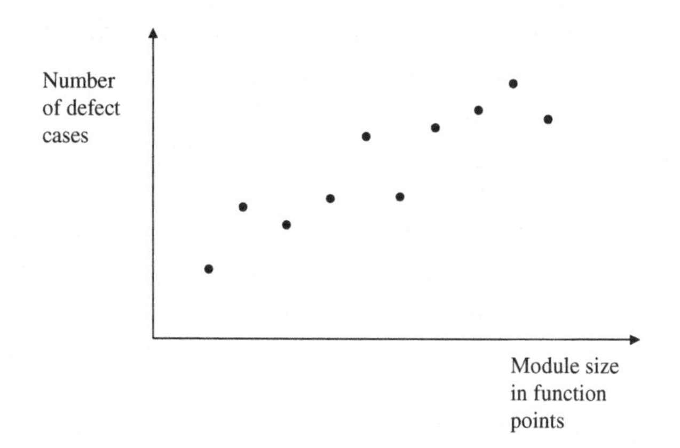

Figure 8.6. SPENTFUEL scatter diagram example.

- Scatter Diagrams (see example in Figure 8.6)

These involve the plotting of two variables to observe any pattern of relationship.

8.4.1. Reviews, inspections and walk-throughs

Throughout the course of this book, we stressed repeatedly that the primary goal of the project manager (PM) is to complete the execution of the work breakdown structures (WBSs). When the work is done, we create a deliverable that is given to the client. But before that happens, we have to ensure that the deliverable has the appropriate quality. How do we do that?

This is a problem that plagues any professional activity. Consider a brain surgeon. She is operating on a patient, finds a strange blue cord in the middle of the brain, cuts it out and sad to say, the patient dies. Is that the end of the matter? No! What happens next is that the surgeon will be invited to a review of the activity with her colleagues. They may not be brain surgeons themselves, but they would be people who could understand the technical processes involved. She would then go through the operation, explain what she had done, and entertain questions from the other surgeons. They would point out that pulling out the strange blue thing was probably not a good idea. She would agree, and from then on, any surgery that took place would not countenance the removal of any strange blue things in the brain. That is the idea behind the review or an inspection or walk-through. You get other people of competent caliber to analyze the work that you have done, criticize constructively, and then make suggestions for improvement for the next time you do that operation. It is quite frankly, the only way we can check on the veracity of professional operations.

These professional checkouts can come in three flavors:

- reviews
- inspections
- walk-throughs

Reviews are formal procedures. They require written documentation, the solicitation of unknown reviewers to analyze the work, and a formal mechanism with which to conduct the review. Think of a PhD defense or a book review. These are examples when you would actually do reviews. Examples in software engineering would include reviewing any software that was used in a nuclear facility, in an aircraft, or in any place for the safety of the public might be compromised by failure. An inspection is slightly less formal. Here, a group of like-minded individuals, but not part of the project team, examine the work and constructively criticize it. The third form, the walk-through, is the least efficient of the three. The walk-through is this where other team members who did not do the work, listen while the team members who did do the work explain to the rest of the team the technical work that they have done. Each form has its own place in the project quality process.

Although reviews are highly expensive, take the most time, and are extremely difficult, they are the most effective and have an adequate effectiveness per cost ratio. Walk-throughs are the cheapest to do, are the least effective, take the least amount of time, are easy, but have a barely adequate effectiveness per cost ratio. Inspections lie in the middle. They are moderately expensive, extremely efficient, take a reasonable amount of time, and are moderately difficult, but have by far the best effectiveness per cost ratio. Inspections are normally the best way to proceed in doing quality checking.

If you are doing the review of nuclear software, you would use a formal review process. The formal review process would be used in any situation where the safety of the public was exposed to a dangerous level. You would use walk-throughs when you are pressed for time and had to maintain a schedule and could not afford the extra luxury of inserting higher quality.

8.4.2. Anatomy of an inspection

Let us go through the basic steps that one would execute in doing an inspection. First of all, a piece of work has to be completed. This is the deliverable from the WBS. This piece of work would be presented to the potential leader of the inspection. The leader, first of all, qualifies the document to make sure it is worthy of reviewing. Just because somebody says they have a document that needs to be reviewed, does not mean that it is adequate to the task.

Inspections take a lot of time and time is money. If the product to be checked is of insufficient quality to even be considered to be reviewed, the leader has the obligation to disqualify it because its quality does not justify the additional expense of doing the inspection. If the leader deems the work product to be satisfactory, he calls a kick-off meeting. The inspection team will consist of the leader, a secretary, and several other technical members. The leader will start the kick-off meeting by disseminating the following information to the team members.

- Familiarize the checkers with the tasks that they have to do.
- Assign role to each so they can role-play a specific function, such as the User, the Accountant, the Quality guy, and so on.

Then the leader hands off the materials to be inspected. This will depend on what the nature of the deliverable of the work breakdown structure was. He will then hand out a schedule of when they will meet and set improvement targets that they hope to achieve. Improvement targets are things like "let us get the number of defects per page up from 5 to 10". Team members will then disperse and analyze the deliverable material.

The third phase of the inspection is the checking phase. Here the individual team members do their own checking. They have to make sure they get it done in time for the logging meeting. The leader will have given them a checklist as a guide to analyze the material. You must make every effort as a checker to follow the rules. Parenthetically, we know inspection itself is a process and must be inspected just like any other processes. The fourth phase is the logging meeting. The purpose of the logging meeting is to record all identified defects. It also is to encourage synergy in the defect detection process. This is something you can see coming out of thesis defenses. One examiner will uncover an error or a discrepancy in the work being inspected. Then the other examiners will immediately pounce in, like sharks swimming in a bloody pool of water, finding more defects by the combined efforts of all the examiners, more than each could find working alone. The idea behind this is not to embarrass the originators of the work; it is to get as many defects out of the work as possible. Remember our quality approach is that the worker cannot make a mistake; it is the process which is broken that permits the worker to make a mistake. During the meeting, the leader leads the group. The secretary records all important information. There is no discussion on the correction of defects; that is not the purpose of the inspection team. The meeting itself should normally be no longer than two hours. Properly done, inspection takes an enormous amount of physical and mental activity. After two hours, you are exhausted and are apt to become unfocused and hence, weaken the inspection process. You want to classify the defects into particular categories. A useful, quick, and effective categorization for defects is the one used by the Public Service to rank software defects into four categories:

A SEV 0 defect is a catastrophic defect. The SEV 0 defect causes the satellite to come out of the sky or the plane to crash. A SEV 1 defect causes very serious repercussions in the execution of the product. Apollo 13 was an excellent example of a SEV 1 defect. It almost caused the astronauts their lives, but they managed to jury-rig the spacecraft and get it back to Earth. A SEV 2 defect is a defect which causes significant impairment in the execution of the product. But it is easy to do a work around and avoid that particular piece of code. Finally, a SEV 3 defect is a trivial defect, a cosmetic defect, a spelling mistake, a colour change, etc., something that is for sure a defect, but has inconsequential effects to the operation of the product itself. Of course, what we have described here is yet another example of a Likert metric being applied to defect classification.

When the logging meeting is finished, the leader and the secretary edit the materials and then do a follow-up with the rest of the team members. The editor cleans up the logs, classifies issues into defects, and categorizes the defects in terms of the SEV ratings. The editor may suggest improvements in the rules. The follow-up is done by the leader. She checks after some time to make sure all of the listed issues have been corrected in writing. She checks that the improvement suggestions have been sent to the appropriate process owners. Finally, she reports all inspection metrics such as the time taken to do the work, the number of defects found, the categorization of the defects, etc.

We conclude this section by giving some very approximate timings for estimating the time to do inspections. It is important to understand that inspections cost time and money. They are WBSs themselves and the input to the inspection is the work product to be inspected and the outputs are the improvement suggestions and the metrics taken in the inspection plus the binary decision to accept or not accept the product. Accepting the product means that that particular work package "earns" its Earned Value as specified in the WBS dictionary.

Inspection timings are approximately:

	Preparation	Meeting
requirements	25 pages per hour	12 pages per hour
functional specification	45 pages per hour	15 pages per hour
logic, design specification	50 pages per hour	20 pages per hour
source code	150 LOCs per hour	75 LOCs per hour
user documentation	35 pages per hour	20 pages per hour

8.5. Quality Closing

Here we compare the final quality metrics against our quality baseline. Understand why certain metrics were not met — were they unrealistic? What should be done in the next project?

8.6. Cause and Effect Diagram Exercise

You are the PM of the SPENTFUEL Control Module development. At the end of one of the system test cycles, you concluded that the quality of the product was quite poor. Your test team reported that the product crashed on average once everyday. The Run Chart showed that the defect backlog was climbing.

When you looked back, you remembered that the development schedule was indeed very tight. The project management plan was written by you and might not have been read by the team. Design review was signed-off before every team member had a chance to review the document. Code review hardly ever happened and all kinds of problems showed up since system integration test, which in turn led to the above situation in system test. When you plotted the defects using the Pareto Chart, you saw that over 50% of the defects related to lack of exception handling and 30% of the defects related to memory leaks. Another fact you were facing was the turnover of 20% of your staff throughout the project, either through internal rotation or attrition.

Your director requests that before you start the next project, you derive at least three improvement initiatives as corrective actions, so that history will not repeat itself. Use the Cause and Effect Diagram to perform the analysis and in point form, suggest the improvement initiatives to your director (with reasons).

8.7. Quality Tools and Techniques Exercises

We presented some traditional quality tools and techniques in this chapter. Research and describe the following modern quality tools and techniques as exercises.

- Affinity Diagram
- Force Field Analysis
- Interrelationship Digraph
- Matrix Diagram
- Nominal Group Technique
- Prioritization Matrix
- Process Decision Program Charts
- Tree Diagram

8.8. Multiple Choice Questions

1. Which project management process does Perform Quality Assurance under Project Quality Management belong to?
   (a) Initiating
   (b) Planning
   (c) Executing
   (d) Monitoring and controlling

2. Your team just finished a final round of regression test on a product. Five minor defects were found. The team worked hard to fix the defects and re-tested the product (with defect fixes) just in time for release to the customer next Monday. What cost of quality applies to the effort spent by the team in fixing those defects?
   (a) Prevention
   (b) Appraisal
   (c) Internal failure
   (d) External failure

3. Which of the following is NOT a project quality planning tool/technique:
   (a) Design of experiment
   (b) Cost of quality
   (c) Quality management plan
   (d) Flow chart

4. As part of the quality audit, the scope statement is checked against the work results to ensure conformance to the customer requirements. The results should be documented and used for:
   (a) Estimating future projects
   (b) Changing the future scope
   (c) Defining future project tasks
   (d) Validating the quality process

5. Pareto analysis, cause and effect, and flow charts are all tools used to:
   (a) Perform quality control
   (b) Perform benchmarking
   (c) Plan quality
   (d) Verify quality

References

1. Garvin, D. A. What Does 'Product Quality' Really Mean? *MIT Sloan Management Review* 26, No. 1, Fall 1984.
2. Paulk, M. C., Weber, C. V., Curtis, B. and Chrissis, M. B. The Capability Maturity Model: Guidelines for Improving the Software Process. Carnegie Mellon University, Software Engineering Institute, Boston: Addison Wesley, 2003.
3. Deming, W. E. *Out of the Crisis.* Cambridge, MA: MIT, 1986.
4. Juran, J. M. and Gryna, F. M. Jr. *Quality Planning and Analysis: From Product Development Through Use.* New York: McGraw-Hill, 1970.
5. Crosby, P. B. *Quality is Free: The Art of Making Quality Certain.* New York: McGraw-Hill, 1979.
6. Pressman, R. *Software Engineering: A Practitioner's Approach*, 7th Edn. McGraw-Hill, 2010.
7. ISO/IEC 9126–1:2001 Software engineering — Product quality - Part 1: Quality model. International Organization for Standardization, Geneva, 2001.
8. Project Management Institute. *A Guide to Project Management Body of Knowledge* (PMBOK), 5th Edn. PMI, 2013, p. 235.
9. IEEE Standard 730–2002. Software Quality Assurance Plans. IEEE, 2002.
10. ISO 9000:2000 Quality Management Systems (0–4). International Organization for Standardization, Geneva, 2000.
11. Ishikawa, K. *What is Total Quality Control? The Japanese Way.* Englewood Cliffs, NJ: Prentice-Hall, 1985.
12. Ishikawa, K. *Guide to Quality Control.* White Plains, NY: Quality Assurance, 1989.

CHAPTER NINE

Human Resource Management

Tom DeMarco once famously wrote the three most important factors in successful project management, are People, People, and People. Having the right selection of people in the project will forgive a myriad of technical failures. Having the wrong people will guarantee the project will fail regardless of how confident you are technically. As Charlton Heston famously said in the movie *Soylent Green*, "People! They're eating people!" Bad projects will do that to people. The purpose of this chapter, then, is to ensure that we have the appropriate selection of people on the team, that we monitor the team's execution, and that we take steps to improve the team's execution if we sense that something is wrong. We will talk about human resource planning and execution, and most importantly, human resource monitoring and controlling. We will also examine various theories of behavioural analysis that may help improve the interactions of our team members. We will conclude by saying a few words on right-sizing and downsizing.

9.1. Human Resource Initiating

The project manager (PM) is identified in the project initiating phase. The PM in turn has to identify the project sponsor, stakeholders, customers, end users, etc. When the PM writes the charter, she will identify the major roles of team members that will be needed on the project.

9.2. Human Resource Planning

In this phase, the project roles, responsibilities and reporting relationships are identified and documented. The clearest way to do this is with an organization chart. An example of the organization chart for the SPENTFUEL Control Module is shown in Figure 9.1. This is part of the deliverables of the Human Resource Plan WBS expansion in Figure 5.3(i).

The human resource plan forms part of the PMP. The responsibility assignment matrix (RAM) follows the RACI format outlining who is Responsible, Accountable, Consulted, or Informed for each assignment. Again, the Human Resource Plan WBS expansion in Figure 5.3(i) delivers the example in Table 9.1.

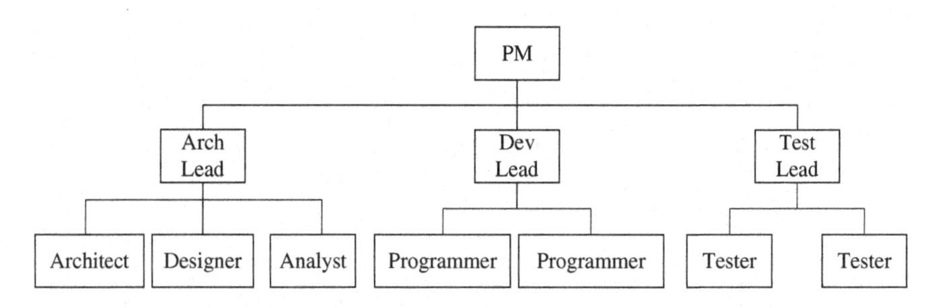

Figure 9.1. SPENTFUEL control module organization chart example.

Table 9.1. SPENTFUEL control module responsibility assignment matrix example (RACI format).

RAM	PM	Architect	Designer	Programmer	Tester
PMP	R,A	C	C	I	I
Architecture	A	R	I		I
Design	A	C	R	I	
Code	A		C	R	
Testing	A	C			R

9.3. Human Resource Executing

In this phase, project staff is acquired and the project team is developed. By "developed", we mean the actions of the team members are continuously monitored and team interactions are improved throughout the execution of the project.

9.3.1. Acquire project team

Acquisition of project staff is usually done via pre-assignment, negotiation, acquisition (hiring, subcontracting, etc.), or formation of a virtual team (i.e., team not physically co-located).

Figure 9.2 is an illustration of the Tuckman–Jensen model [1] of team formation. During the team's life cycle, there will typically be five phases: forming, storming, norming, performing, and adjourning. In the forming phase, there is considerable uncertainty. Team members will be meeting each other for the first time. They will have no idea of the strengths, weaknesses, and particular communication characteristics of each other. In the storming phase, team members start to test and challenge each other. In the norming phase, they open up and exchange ideas. In the performing phase, the team works consistently and effectively. Finally, in the

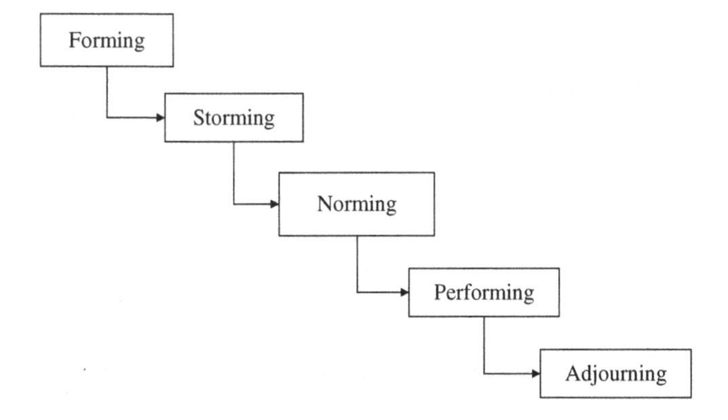

Figure 9.2. Team assembly.

adjourning phase, there is a sense of loss due to separation and team anxiety goes up. Will there be a new project? When will a new team be formed? As leader of the team, the PM has to celebrate the accomplishments of the project and prepare the team members for their next adventure.

9.3.2. Develop project team

To develop the project team, competencies and interaction of team members must be improved. In addition to training that improves technical competencies of the team members, there should also be interpersonal (and other soft skills) training. Team-building activities should also be organized. Upon discovery of values that are important to one another, ground rules can be established. For example, those who are late for team meetings could be fined $1 and when sufficient funds have been collected, the team might go for a Friday afternoon celebration at the local watering hole. Team development will also benefit from physical co-location and appropriate reward and recognition.

During project execution, there will always be changes. Figure 9.3 illustrates the phases that the team will go through when changes are made to the team's structure or operations.

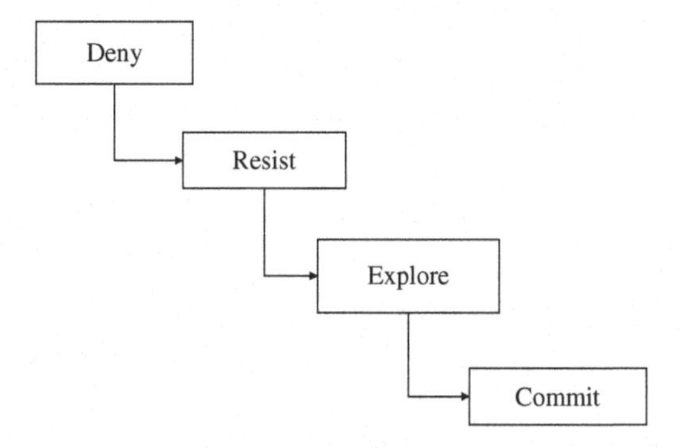

Figure 9.3. Adjusting to changes.

- Deny — the PM has to inform the group of the desired change, and conduct one-on-one interviews with the individuals that are directly involved in the change.
- Resist — all humans resist change. The PM has to show empathy, realize that resistance is normal, and continue to encourage the team to incorporate the change.
- Explore — the PM has to facilitate networking and provide feedback.
- Commit — the team has incorporated the change. The PM has to publicly recognize the adjustment accomplished and give appropriate rewards.

9.3.3. Motivation theories

There will come a time in every PM's life when the team seems unmotivated; how can the PM turn this around? Here we discuss three motivational theories that psychologists have developed.

9.3.3.1. *Maslow's hierarchy of needs*

Alexander Maslow [2] realized that humans have distinct needs. He arranged them in a hierarchy, as shown in Figure 9.4. The five

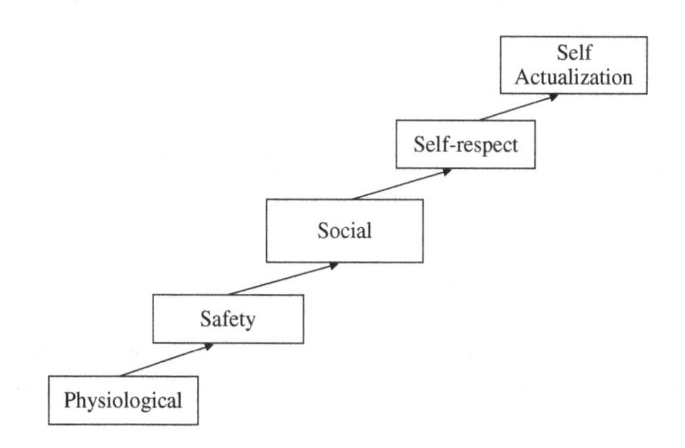

Figure 9.4. Maslow's hierarchy of needs.

levels of needs begin with the physiological need (is there enough food?), then safety (safe from danger and threats), social (there is love and friendship), self-respect (self-esteem), and self-actualization (be able to use one's creativity, develop skills, and possess power). Maslow's basic insights in this hierarchy are that the lowest level has to be satisfied before the human can attempt to satisfy a need at the next higher level.

9.3.3.2. *McGregor's theory X and Y*

Theory X [3] espouses the 19th century view that workers are stupid, lazy, and have no ambition. Therefore strong dictatorial management is needed to keep the workers in line. Theory Y [3] espouses the opposite view that people are eager and willing to accept responsibilities, as well as directing their own effort toward organizational objectives.

9.3.3.3. *Herzberg's theory of motivation*

Herzberg's Theory of Motivation [4] claims that poor hygiene factors such as low pay, unsafe work conditions, bad attitude of supervisors, etc., will demotivate staff. By improving these hygiene factors, team motivation will improve too, at least for a short period of time. The real team motivators are personal growth and the opportunity to achieve self-actualization.

9.4. Human Resource Monitoring and Controlling

This phase focuses in tracking team member performance, providing feedback, resolving issues, and coordinating changes to enhance project performance. Clear objectives have to be set for individuals. Recall that the five most important characteristics of objectives can be described by the acronym SMART (Specific, Measurable, Agreed to, Realistic, Time-bound).

9.4.1. Management style and power

Our first objective is to have competent management in the running of the project. Several management styles can be identified to establish direction, align, motivate and inspire people.

- autocratic (i.e., dictatorial)
- laissez-faire (i.e., hands-off)
- democratic (i.e., participative)

There are various types of power that the PM could use. Here are a few:

- legitimate (derived from formal position — e.g., VP of Procurement has purchasing authority to a certain limit)
- coercive (manage by fear)
- reward
- expert (earn others' respect from own expertise)
- referent (cite the authority of a powerful person)
- purse string (who has the money?)
- bureaucratic (use rules and procedures in organization to benefit oneself or one's group, and block others from gaining such benefit)
- charismatic (encourage others to do things they may not be inclined to do — say this to your child tonight, "good kids go to bed at 9")
- penalty

9.4.2. Conflict management

The next major objective is to reduce conflict management. The top sources of conflict include schedule, project priorities, personnel resources, technical opinion, technical performance, administrative procedures trade off, cost, personalities, etc. The traditional view is that conflict is bad and created by troublemakers. It should be avoided and suppressed at all costs. The contemporary view is that conflict is inevitable, beneficial, and going to result in good changes, so embrace it!

The Thomas–Kilmann model [5] looks at various modes of handling conflicts by plotting them on the assertiveness and cooperativeness axes as shown in Figure 9.5.

- Confronting (or problem solving): concerns from both sides are equally important and tabled for discussion, resulting in a win-win resolution.
- Compromising: partially satisfying both sides, but not really getting what they want, resulting in a lose-lose resolution.
- Smoothing: de-emphasize differences and preserve harmony — a weak mode of handling conflict.
- Withdrawing: break the connection for both sides to calm down and come back with additional data.
- Forcing: one side wins and the other loses resulting in a win-lose resolution — typically used to resolve critical issues.

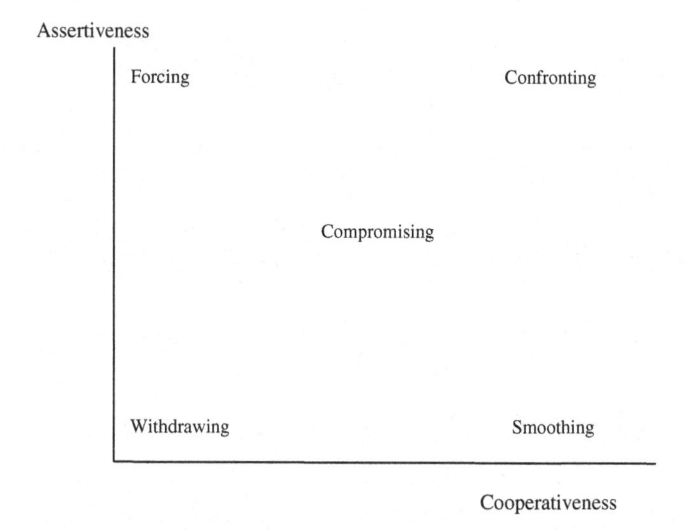

Figure 9.5. Thomas–Kilmann conflict mode instrument.

9.5. Human Resource Closing

Archive project team performance appraisal and issue logs. Document lessons learned.

9.6. Behavioural Analysis

The following sections outline a few methodologies for team and individual behavioural analysis that will help significantly in human resource management. Note that there are no rights or wrongs to the results of analysis. No two persons have the same behavioural characteristics. That is why people have to cope with diversity. It is important for people to understand their own characteristics and learn how to co-exist with others.

9.6.1. Blanchard's situational leadership

Ken Blanchard [6] developed Situational Leadership with Paul Hersey at Ohio University in 1968. Situational Leadership II was refined based on the thinking and research of colleagues, and feedback from users. The results of assessment, based on 20 questions, reveal flexibility and effectiveness in using styles of leadership.

In brief, people want to perform well. It is the job of the leader to bring out that magnificence in people and to create an environment where they feel safe and supported, and do the best job possible in accomplishing key goals. Today, leaders must be partners with their team. They can no longer lead from positional authority.

In Figure 9.6, the x-axis is the Directive Behaviour such as teaching, supervising, structuring, and organizing. The y-axis is the

Supportive Behaviour

D3: capable cautious performer *Var: commitment* M-H: competence ***S3: Supporting***	**D2: disillusioned learner** *L: commitment* L-M: competence ***S2: Coaching***
D4: self-reliant achiever *H: commitment* H: competence ***S4: Delegating***	**D1: enthusiastic beginner** *H: commitment* L: competence ***S1: Directing***

Directive Behaviour

Figure 9.6. Blanchard model.

Supportive Behaviour such as encouraging, listening, asking, and explaining. S1–S4 are styles of leadership exhibited by the leader. D1–D4 are development level exhibited by the team members with commitment being a combination of motivation and confidence, and competence with goal or task-specific skills and knowledge.

- In the first quadrant (D1/S1), the team members (probably just joining a project) are enthusiastic beginners with high commitment and low competence (as they do not know much about the project). The leader must provide project direction.
- In the second quadrant (D2/S2), the team members are disillusioned learners with low commitment (opened a can of worms!) and low to medium competence (as they are learning more about the project). The leader needs to provide coaching.
- In the third quadrant (D3/S3), the team members are capable, yet cautious performers with various commitments (depending on how many actually align with the project mission and vision) and medium to high competence. The leader needs to provide support. Examples might be: arrange for project-wide training, buy sufficient software licenses for the team, and so on.
- In the fourth quadrant (D4/S4), team members are self-reliant achievers with high commitment and competence. The leader needs to delegate work instead of micromanaging.

Overall, the Blanchard Model measures whether the team leader recognizes which quadrant each team member is situated in. He also needs to know the quadrant that the whole team is in. The model also measures whether the team leader is flexible enough in his involvement with the project (hands-on or hands-off at the right time).

9.6.2. Fundamental interpersonal relationship orientation behaviour (FIRO-B)

FIRO-B, invented by Dr. Will Schutz [7], attempts to analyze how team members interact with each other. It does this by giving team members a 15-minute survey, measuring the levels of each individual's needs along three interpersonal dimensions: inclusion, control, and affection. The survey consists of 54 questions, each with 6 ratings. The methodology examines each interpersonal dimension, analyzing whether the team member wants to express or give, or does one exhibit want or receive behaviour. Specifically, it examines:

- Inclusion (invite versus participate)
- Control (direct action versus take order)
- Affection (reach-out versus share)

It is useful in helping individuals understand how they relate to other team members and how they can improve that interaction. FIRO-B is often combined with Myers–Briggs (FIRO-B categories are mutable while Myers–Briggs are absolute and unchangeable).

9.6.3. Myers–Briggs type indicator (MBTI)

Katherine Cook Briggs and her daughter Isabel Briggs Myers were disciplined observers of human personality differences and were strongly influenced by the great Swiss psychiatrist, Carl G. Jung. Myers–Briggs [8] is arguably the most commonly used methodology for behavioural analysis. It examines people from four dimensions as described below:

- Extraversion (E) versus Introversion (I): focuses on outer world and people versus inner world of ideas and impressions
- Sensing (S) versus Intuition (N): experience and practical versus imaginative and seeing the big picture
- Thinking (T) versus Feeling (F): logical and rational versus preserving harmony
- Judging (J) versus Perceiving (P): focuses on deciding and finishing versus being flexible and adaptable

Team members can use questionnaires, or just self-categorization, to come up with their "type". There are 16 "types", such as

ESTJ. Understanding each other's "type" can improve team interactions.

One has to watch out for extreme scores for these personality types. For example, too strong a sensing personality will miss the big picture. Some people cannot provide effort estimates necessary to code a software module until sufficient experience and practice has been gained, by which time the project would have been almost done, rendering the estimates irrelevant. Conversely, too strong an intuition personality may lead to setting unrealistic goals and a constant swapping of goals upon discovery of practical problems exposed by actual experience. Too strong a thinking personality will dismiss the feelings of those affected by certain decisions even though the logic and rationale fully support the decision. Too strong a feeling personality will lead to decisions being made due to personal preference or likeness instead of sound reasoning.

9.6.4. Parker's team player survey

Glenn Parker [9] developed his methodology to identify one's style as a team player by completing 18 sentences with possible endings ranked in order of applicability. The four styles all begin with the letter "C":

- Contributor

  Individual who focuses on details and data. Too strong a contributing style may miss the big picture.
- Collaborator

  Individual who is goal directed and forward looking; however, too strong a collaborator style does not confront.
- Communicator

  Individual who is an effective listener and promotes a positive climate; but will not focus.
- Challenger

  Individual who is candid, assertive, and willing to disagree. The drawback is that one may push too far and may not back off.

9.6.5. Porter

Dr. Elias Porter uses a model called the Strength Deployment Inventory (see Figure 9.7) [10] to help people identify the personal strengths in relating to others under two conditions: when everything is going well and when they are faced with conflicts (or pushed to a corner). The questionnaire has 20 sentences each with 3 endings. The answerer distributes 10 points across the 3 ways. Red, blue, and green are the colours to represent the personal strengths that all begin with the letter "A".

- Red (Assertive directing): an individual who persuades, challenges, and takes risks.
- Green (Analytic autonomizing): an individual who is a thinker, cautious, thorough, self-reliant, and self-sufficient; and does things in a meaningful order.
- Blue (Altruistic nurturing): an individual who promotes harmony, makes life easier for others, and is in general warm-hearted.

It is also possible for people to possess combinations of personal strengths. For example, red can combine with green, green with blue, blue with red. In particular, flexible cohering describes individuals who exhibit all three personal strengths.

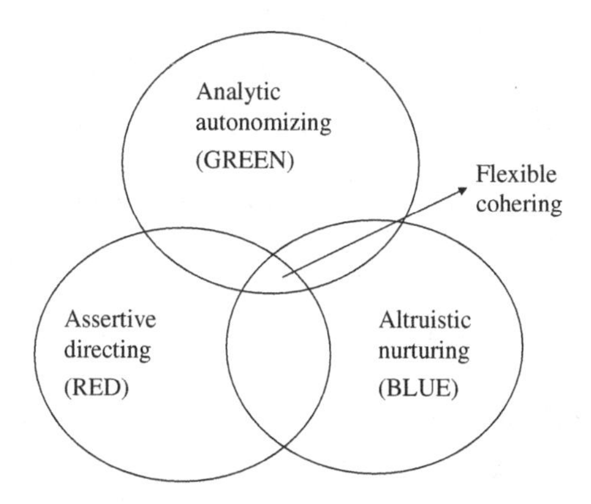

Figure 9.7. Strength deployment inventory.

9.7. Right-Sizing and Downsizing

In today's world, corporations are all striving to do more at lower cost. With offshoring and outsourcing (see Chapter 12) as common practice, in-house right-sizing is needed through downsizing. PMs may be directly or indirectly responsible for downsizing to management-specified targets, depending on the organizational structure (i.e., functional, matrix, projectized). Some typical mechanisms to achieve right-sizing include separation with bottom performers and separation with wrong skill-sets. For separation with bottom performers, each department, sector, or division would simply remove the bottom, say 5% or 10% of its staff to attain the right-sizing goal. During quarterly, bi-annual, or annual performance reviews, staff will be categorized into the top 20%, mid 70%, and bottom 10%, for example. After the removal of the 10% bottom performers, the next wave of right-sizing may occur. Now the resource managers have to name the next bottom 10% from the remaining workforce. How can this be done? Is there an end to this? Some resource managers simply say, "I can't do it anymore!" But this is part of life, so we must deal with it in an ethical, fair, and professional manner.

Here is a simplified example. Your child hosts a pool party in your backyard inviting 9 other friends. You decide to come home with 10 gifts for the kids. Surprise! When you enter the backyard, you see more than 20 kids. How would you distribute the gifts? How about the first 10 who swim from one end of the pool to the other end will get a gift? You think this is a good idea until you realize that some do not know how to swim. Then, you keep thinking — how about having the kids run from one side of the fence to the opposite end and the first 10 will get a gift? You soon realize that two of the kids are actually in wheelchairs. Finding an equitable solution to this puzzle is not easy.

Similarly, executives are thinking about how to grow the business with less people doing more work every single minute. What criteria should be put in place to reward the top contributors and team players, and to separate from the less valued contributors with fairness?

Separating people with the wrong skill-sets is less rational but it can be done much faster. The whole department, sector, or division can simply be separated, sold to another company, or spun off as a different company. Some organizations will offer staff opportunities for re-training (for example, from structured programming to object-oriented, from testing to programming). Note that if individuals are not flexible and adaptable to changes, the only alternative left is separation.

Separation without cause (or job no longer justified in-house) is typically compensated with a severance package in accordance with (or mostly exceeding) government regulations. Of course, dismissal (separation with cause) applies always and immediately to unethical behaviour (e.g., discrimination, sexual harassment, stealing/fighting/drunkenness at work, etc.) without compensation. See Chapter 14 for additional details.

9.8. Behavioural Analysis Exercise

You are the PM of a sub-function of the SPENTFUEL Control Module. The team was formed not long ago. You kicked off a meeting on Monday morning to discuss the project plan. Here's part of the discussion:

You: Good morning, how's everybody today?
Alan: Fine.
Dave: Alright.
Pat: Great! I had a wonderful weekend doing .... (and on and on)
Sam: (*silent*)

You: Let's talk about the project plan this morning. I have worked very hard on this over last week. It should be perfect, so I'll just run it by you.

Dave: Why didn't you engage us in the work?

You: Well, the team is new and I have been involved in some background activities for a few weeks. So, I went ahead with the project plan.

(Later on when you talked about scope...)

Dave: Why was Feature A included in the scope instead of Feature B?

(You then explained how CNA and DND made such a decision.)

Dave: But Feature B is very important, you should exercise your influence to make them accept it.

Pat: Well, let's not argue over the past. I am sure that the project is still going to make money for the company.

(Dave was still stressing the importance of Feature B.)

Alan: I think Feature B should not be included in the scope because it requires several pre-requisites that will be expensive to implement and poses several risks to the project.

(Alan then started explaining in detail the specifics and Dave reluctantly shut up.)

(Later on, when you talked about time management...)

Dave: I cannot accept this preliminary schedule. It is too tight and unrealistic.

Pat: I am sure that we can work towards the deadline if overtime pay is pre-approved.

(After a while...)

Alan: I have done some detailed breakdown of the tasks assigned to me. I suggest the following changes and here are the reasons.

(Alan then went on and on. You had no reason not to change the schedule, yet you'll miss the deadline.)

(Later on when you talked about cost management...)

Pat: Oh! You've got to allocate some budget for team building activities. Back in the good days, we'll fly everyone to a resort and spend a few days there. We also need to conduct interim checkpoints with the ADMs of CNA and DND, parliament committee, etc. Invite them to come to our site, show them the prototype and lab set up, and run some demos. You need to include their entertainment in the budget.

(*You have run out of time for the meeting. Sam never said a word and left the room without even making an eye contact with anyone.*)

Use Blanchard, FIRO-B, Myers-Briggs, Parker, and Porter to analyze the behaviour of yourself, your team members, and the team as a whole. Formulate an action plan for your next steps.

9.9. Multiple Choice Questions

1. The PM is selected for a project during:
   (a) Initiation
   (b) Planning
   (c) Execution
   (d) Termination

2. Which project management process does Acquire Project Team under Project Human Resource Management belong to?
   (a) Initiation
   (b) Planning
   (c) Execution
   (d) Control

3. Which of the following is true for Herzberg's Theory of Motivation?
   (a) Poor hygiene is inevitable
   (b) Poor hygiene means low pay only
   (c) Improve hygiene may not improve motivation
   (d) Improve hygiene lead to opportunity to achieve self-actualization

4. Your PM has a relaxed attitude and stays away from making decisions for the project. What management style is this?
   (a) Autocratic
   (b) Laissez-faire
   (c) Democratic
   (d) Participative

5. Which of the following is not part of team assembly?
   (a) Resisting
   (b) Adjourning
   (c) Storming
   (d) Performing

6. A project loses a team of contractors in the middle of a project. A new project team is formed to replace the old team. As the PM, what is the first topic to address to the team in the kick-off meeting?
   (a) Identify team roles and responsibilities
   (b) Review detailed schedule
   (c) Discuss cost estimates
   (d) Emphasize your authority

7. Constructive team roles include:
   (a) Encourager, initiator, and gatekeeper
   (b) Information giver, devil's advocate, and clarifier
   (c) Withdrawer, harmonizer, and blocker
   (d) Summarizer, recognition seeker, and information seeker

8. Of the following conflict management approaches, which is believed to lead to the least enduring positive results?
   (a) Problem solving
   (b) Avoidance
   (c) Compromise
   (d) Forcing

9. Conflict resolution techniques that may be used on a project include:
   (a) Withdrawing, compromising, controlling, and forcing
   (b) Controlling, forcing, smoothing, and withdrawing
   (c) Confronting, compromising, smoothing, and directing
   (d) Smoothing, confronting, forcing, and withdrawing

10. A likely result of using "compromise" to resolve a two-party conflict is:
    (a) Lose-lose
    (b) Win-lose
    (c) Win-win
    (d) Lose-win

References

1. Tuckman, B. W. and Jensen, M. A. C. Stages of small group development revisited. *Group and Organizational Studies*, 2, 1972, pp. 419–427.
2. Maslow, A. *Motivation and Personality*. New York: Wiley, 1954.
3. McGregor, D. *The Human Side of Enterprise*. McGraw-Hill, 1960, 2005.
4. Herzberg, F. B. *et al. The Motivation to Work*, 2nd Edn. Wiley, 1959.
5. Available at: http://en.wikipedia.org/wiki/Thomas_Kilmann_Conflict_Mode_Instrument.
6. Blanchard, K., Carew, D. and Parisi-Carew, E. *The One Minute Manager Builds High Performance Teams*. William Morrow and Company, 1990.
7. Schutz, W. *The Human Element: Productivity, Self-Esteem and the Bottom Line*. Jossey-Bass, 1994.
8. Myers, K. D. and Kirby, L. K. *Introduction to Type Dynamics and Development*. CA: Consulting Psychologists Press, 1994.
9. Parker, G. M. *Cross-Functional Teams*. Jossey-Bass, 1994.
10. Porter, E. H. *Strength Deployment Inventory*. Personal Strength Publishing, 1992.

CHAPTER TEN

Communications Management

There are multiple communication targets for the project manager — the project team, stakeholders, executive sponsors, etc. The purpose of this chapter is to examine the different ways by which we communicate. Communications management includes processes that ensure timely and appropriate generation, collection, dissemination, storage, and archiving of all project information continuously through the execution of the project. What needs to be communicated? Project information can include the following:

- status
- problems
- updated project plans
- successes / failures
- team member performance
- meeting schedule
- new risks
- date of the next milestone

10.1. Introduction to Communications

When should information be communicated? Usually, information can be communicated on a periodic basis. That is, we specify in the project plan, the dates when it should be delivered such as weekly, biweekly, monthly, etc. It can also be communicated on an episodic basis, where some emergency has arisen and information must be provided to deal with the crisis.

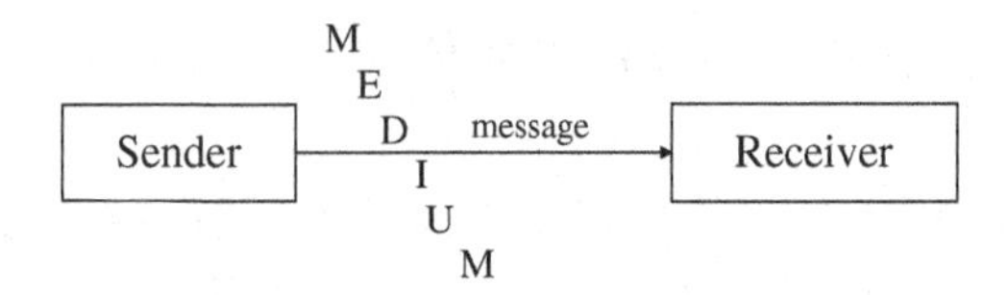

Figure 10.1. Generic communication topology.

Figure 10.1 illustrates a very simple communications model. The sender wants to send a message to the receiver over a particular medium. There are various modalities that we can use in transmitting our information. Of course, we normally use verbal or written communication. It is also important to understand that most of our face-to-face communication is in fact done in a nonverbal sense; the tensing of shoulders, the furrowing of eyebrows, things like that impart far more information than just the words spoken. Psychologists estimate that 70% of our face-to-face communication is in fact done in a nonverbal mode. Other modalities include paralingual modalities where we use nonverbal expressions and shading in our speaking by adopting a shrill or conciliatory tone of voice. Listening is as important as speaking. It is important to understand that the receiver must be involved in both active and effective listening. There is a reason the Lord gave us two ears and only one mouth! It is also important in communication that the receiver gives constant feedback to the sender to indicate that the message has been received and understood.

There are several dimensions of communications that can be used. Communications can be written or verbal. It can be internal or external. It can be formal or informal. It can be sent vertically up to supervisors and down to subordinates within the organization or horizontally to peer groups. Face-to-face communication as we mentioned can also be verbal or nonverbal.

In Figure 10.1, there is only one communication channel. For a team size of n, there are $n \times (n - 1)/2$ communication channels. In other words, the number of channels is proportional to the square of the number of users! It is desirable to break larger team into smaller/core teams with an optimal size of 4–9.

10.2. Communications Planning

The information needed by stakeholders and the communication vehicle need to be determined. The communications management plan consists of:

- communication item
- stakeholder
- purpose
- frequency
- start/end date
- format/medium
- responsibility

Inputs to communications planning include specifying the types of technologies that can be used such as, but not limited to:

- brief conversations
- extended meetings
- simple emails
- written documents
- online databases

Factors that we need to include in the communications planning include:

- immediacy of need for information
- availability of technology
- expected ability of people
- length of the project

We have already identified the stakeholders in the Stakeholder Registry (see Section 13.1). We need to understand and elaborate the communications needs and characteristics of those stakeholders. In doing communications planning, we must adopt a need-to-know basis and conserve resources whenever we can. Communication is an overhead. Do a benefit-cost analysis of the communications strategy. Plan on sending the minimum amount of information necessary to keep the stakeholder up to date on the project's progress.

The output from communications planning is the Communications Management Plan. This is the document that describes the details of the communication methods that will be used to gather and store information. The plan will define procedures to collect the specified data. It will also indicate a distribution algorithm describing to whom and when information is to be delivered. It will contain a description of the information to be distributed. It will indicate how stakeholders can access information in between scheduled communication points, and of course, as with all project management plans, there will be a process for improving the plan as the project progresses.

10.3. Communications Executing

Information is provided to stakeholders in a timely manner. The first formal activity in communications execution is a special case. That is the kick-off meeting.

10.3.1. Kick-off meeting

Kick-off meetings are held at the beginning of the project and at the beginning of each phase. The purposes of kick-off meetings are to:

- get the team members to know one another,
- establish working relationship and lines of communication,
- set team goals and objectives,
- review project status and plan,
- identify problem areas,
- establish individual/group responsibilities and accountabilities, and
- obtain individual and group commitments.

10.3.2. Tools and techniques in communications executing

Certain basic communication skills are necessary in communications execution. The sender must make sure that the information sending:

- is clear,
- is unambiguous,
- is complete,
- has been received correctly, and
- has been understood.

The receiver, on his part, has to ensure that the information has been completely received and understood, and communicate that fact to the sender. As with any modality of communication, there are activities that can interfere with effective communication which we call communication blockers. Some examples are:

- channel noise
- distance
- improper encoding of messages
- saying "that's a bad idea"
- hostility
- language
- culture

The communications plan may prescribe the use of information retrieval systems such as:

- manual filing
- electronic databases
- project management software
- web sites

It may also indicate access to documentation, including engineering drawings, designs, specifications, test plans, and so on.

There are many different ways that we can distribute information. The following list is not inclusive. It is intended to give you possible modalities of information distribution:

- project meetings
- document distribution
- web sites
- Fax
- electronic mail
- voice mail
- videoconferencing
- project intranet

In passing, the running of meetings is an extremely important attribute for a PM. There is nothing more frustrating in one's professional life to participate in a poorly run meeting. The amount of time wasted can be enormous. Image a team of 10 in a 1-hour meeting, it translates to 10 person-hours of effort. Here are some useful hints for running productive, useful meetings:

- set a time limit,
- schedule recurring meetings in advance,
- meet regularly but not too so,
- have a purpose for each meeting,
- have an agenda,
- distribute agenda before meeting,
- stick to the agenda,
- let people know in advance their responsibilities,
- bring the right people together,
- assign deliverables and time limits to all tasks assigned, and
- document and publish meeting minutes.

10.3.3. Manage communications

This is the time for the PM to work the plan, since she has already planned the work. To do that, she must manage communication to satisfy the needs and resolve any issues that may have cropped up. This will help limit disruptions during the project and make sure that all related parties are proactively involved in times of project crises.

10.4. Communications Monitoring and Controlling

In this phase, performance will be reported by comparing achievements with plan, and reviewing plan and options against future scenarios. The Stakeholder Registry will indicate the nature and timing of the performance information to be reported. Some of the items to be reported include:

- status report
- progress report
- forecasting report
- trend report
- variance report
- exception report

Note that status report applies to the overall project while progress report applies to an update of a certain item of interest (e.g., fixing a software bug). Forecast can be a subjective prediction of status and progress (e.g., by working overtime, the slippage over the past few weeks will be recovered by next week) versus trend analysis which purely depends on the objective view of where the project is heading based on, for example, some mathematical technique or graphical plotting. Variance analysis reports deviation from the baseline or goals set specifically for the current project. Exception report provides early warning of deviation beyond tolerance levels and highlights variance that exceeds certain predefined threshold (e.g., SPI or CPI deviates over 20% from 1.0). Based on the problems detected, corrective actions should be initiated and further work can be authorized.

10.5. Communications Closing

The PM needs to notify all stakeholders involved in the project of its closure. He also needs to inform the stakeholders of all the final performance indicators as to whether the project was a success or failure and why. Then, the PM has to lead the construction of the lessons learned document.

The purpose of lessons learned is to identify project success or failure, and to recommend follow-on actions for future performance improvement. Inputs should be concerned with all aspects of the project including, but not limited to, technical, managerial, and process. It has been suggested that the PM should be responsible for the lessons learned session. However, there has also been alternative view that no managerial staff be present so that team members are free to voice their opinions and concerns. Inputs can be given on-line prior to the session so that the chair person (project manager or another staff) can sort out the inputs, group inputs into categories, and eliminate duplicates. Inputs can also be taken during the lessons learned meeting, in which case, it is suggested that a round-table, token-passing process be followed. This will avoid opinions only being heard from the outspoken individuals and everybody will be given a chance to speak up. When opinions are initially gathered, all inputs count and no idea is a bad idea. In the subsequent phase, inputs are sorted based on priority and actions are derived to tackle the top 3–5 issues and bring improvements. The results are not limited to update or improvement of:

- lessons learned knowledge base
- corporate policies, procedures and processes
- products or services
- business skill

Note that the lessons learned document should also capture what went well in the project so that good practices will be preserved and continued.

While reading a book is boring, an extraction from SARK[1] is added below as an interesting example for lessons learned.

- Lesson One
  I walk down the street
  There is a deep hole in the sidewalk
  I fall in
  I am lost ... I am helpless
  It isn't my fault. It takes me forever to find my way out.

- Lesson Two
  I walk down the street
  There is a deep hole in the sidewalk
  I pretend I don't see it
  I fall in again
  I can't believe I'm in the same place
  But it isn't my fault
  It still takes me a long time to get out

- Lesson Three
  I walk down the street
  There is a deep hole in the sidewalk
  I see it is there
  I still fall in ... it is a habit
  My eyes are open
  I know where I am
  It is my fault
  I get out immediately

- Lesson Four
  I walk down the street
  There is a deep hole in the sidewalk
  I walk around it

- Lesson Five
  I walk down another street
  Hopefully, there is no Lesson Five (b) whereby "I fall into another deep hole".

10.6. Exercises

What is the best form for each of the following, if we measure them along two dimensions, written or verbal (W or V), or formal or informal (F or I)?

- Memos
- Project plans
- Communicating over long distances
- Meetings
- Presentations
- Conversations
- Complex plans
- Emails
- Notes
- Speeches
- Project Charter

10.7. Multiple Choice Questions

1. If there are five people in the team, how many communication channels are there?
   (a) 5
   (b) 10
   (c) 15
   (d) 20

2. You used some mathematical means to analyze the current project metrics and use that to predict the future project status. This is known as:
   (a) Forecast
   (b) Trend analysis
   (c) Variance analysis
   (d) Exception analysis

3. Ideally, communication between the project manager and team members should take place:
   (a) Via daily status reports
   (b) Through approved documented forms
   (c) By oral and written communication
   (d) Through formal chain of command

4. Effective ways to manage stakeholder expectations include all of the following project elements except:
   (a) Clear requirements definition
   (b) Control scope change
   (c) Timely status information
   (d) Frequent cost reports

5. What is the most effective process to ensure that cultural and ethical differences do not impede success of your multi-national projects?
   (a) Co-locating
   (b) Training
   (c) Forming
   (d) Teaming

References

1. SARK. *Inspiration Sandwich: Stories to Inspire Creative Freedom.* Celestial Arts, 1992.

CHAPTER ELEVEN

Risk Management

Being a project manager (PM) in the ancient world was a much easier proposition than it is today. The ancients believed that the Gods controlled every aspect of human life and that individuals have no control over their destiny. Therefore, when the Romans, for example, went to war, the most they would do in terms of battle preparations, would be to open up a chicken and check its entrails to see if the Gods would provide for a victory or not. They knew the Gods had predetermined the outcome of the battle. In a certain sense, that made life a lot easier for the PM. If the project were a success, that was because the Gods had decreed so. If the project failed, that was because the Gods decreed that too; it was no fault of the project manager. In his remarkable book, *Against the Gods*, Peter Bernstein makes the argument that the concept of risk separates the modern world from the ancient. Sometime in the 17th century, French mathematicians began to apply what we now call probability theory to improve their chances of winning in games of chance. For the first time, humans realized they had some control over their chances of success in projects. The rest of this unit describes a remarkable improvement in the execution of projects; the definition and management of risk. We introduce the concept of risk, how we identify risks, and more importantly, how we manage risks.

Good risk quotes:

- "If you know your enemy and yourself, you need not fear the results of 100 battles." Sun Tzu 500 B.C.E. [1]

- "If you do not actively attack risk, risk will actively attack you!" Anonymous
- "Risk prevention is more effective than risk detection." Anonymous
- "If you don't ask for risk information, you're asking for trouble." Anonymous

11.1. Risk Introduction

What precisely is risk? Intuitively we imagine risk to be something that could hurt us; something that may or may not happen. The Software Engineering Institute defines risk as being "the possibility of suffering loss". Lawrence defines risk as "the measure of probability and severity of adverse effects" [2]. Another researcher, Rowe, defines risk to be "the potential for realization of unwanted negative consequences of an event" [2]. PMBOK defines project risk as "an uncertain event or condition that, if it occurs, has a positive or negative effect on one or more project objectives" [3]. PRINCE defines risk as "the chance of exposure to the adverse consequences of future events" [4]. There are two categories of risks: known or unknown. There are two types of risks: business (i.e. inherit in project as part of execution) or pure (unexpected).

If an event is a risk, it involves at least three characteristics. The first is uncertainty; the event is not guaranteed to occur. The second is a loss if the event does occur. The third characteristic is that of temporality; risks vary over time. For example, the risk of having an auto accident increases when it is raining or snowing.

For a risk to be understandable to humans, it must be expressed clearly. It must include a description of the conditions that might lead to a loss. It must include a description of that loss or consequence. It must also include the relevancy of time as to the significance of risk.

The Software Engineering Institute in their seminal work on risk management, "Principles of Continuous Risk Management" [5], defined three fundamental principles that must be followed if risk management is to be successful. They are:

- core
- defining
- sustaining

The Core Principle is Open Communication — people do not care very much for unpleasantness. We try to avoid it whenever we can. In fact, in English, we have an appropriate quote called "Don't shoot the messenger". So the instinctive reaction of any member in the organization, sensing a risk about to eventuate, is to keep quiet. This is something we must overcome in the organization.

Let us give you a practical example from industry. In the mid-1970s, National Cash Register (NCR) was the largest cash register company in the world. They had at least 80% of the market worldwide. At that time, all cash registers were mechanical. NCR was about to bring out the greatest line of mechanical cash registers they had ever produced. Now, at the same time, solid-state electronics was evolving to the point where engineers could put a computer and memory on a tiny chip. Add to that a video display terminal and you had a potential competitor to mechanical cash register. Can you imagine the scenario of the NCR salesperson coming into the Mom and Pop convenience store, trying to sell them a mechanical cash register against the representative from the competitor with his electronic version which was much cheaper and far more featured? It would have been a slam dunk as to which one Mom and Pop are going to select. That year, the launch was disastrous for NCR. Sales plummeted and they had to lay off 75% of their sales workforce. The company almost went bankrupt. Now, here is an excellent example of a technical risk eventuating and destroying your perfectly good traditional product. How could that risk have been missed?

Let us go back over the scenario and postulate about who would have been the first person to have seen the risk eventuating? Not the engineers at NCR, not their executives. It would have been the salesperson who could not make a sale because the competition had so superior a product. Why did the salesperson not pass that information on to his boss? Because, in hierarchies, bad news is discouraged. The salesperson would have covered up the bad news with something like "I'll sell better next week. I am just having an off-week." As the news gets passed up the hierarchy, the bad news gets more and more sugarcoated, and made more and more implausible. By the time it hits the people in the boardroom, they think they have a wonderful product launch, when in fact, disaster is about to take the company down. The solution to this is open communication. Everybody in the organization has to feel comfortable in reporting uncomfortable events. If one does that, the organization has time to do something about it.

The second important principle of continuous risk management is the Defining Principle. The defining principle has three dimensions: forward-looking view, shared product vision, and global perspective. The forward-looking view is that risk is something that happens in the future. If that risk has eventuated, it is no longer a risk: it is a fact! Risk implies that we have time to react to avoid the negative consequences of the risk hurting us. So the forward-looking view is extremely important. The second dimension of the defining principle is the shared product vision. Many products in our company will be affected by the same risk. When one project team identifies and manages a risk, it should share that information with all the other project teams so they do not have to rediscover it on their own. The third dimension of the defining principle is the global perspective. We live in a global world. Events that happen halfway across the world can have dramatic effects on our projects. Innocent risks that one would not think would affect the project include examples like the price of oil, the price of exchange currency, the price of gold, and so on.

The third principle of continuous risk management is Sustaining. Sustaining too has three dimensions: integrated management, teamwork, and continuous process. The integrated management dimension stresses that all of management must be involved in risk assessment. It is something that needs to be permeated through the entire organization. The teamwork dimension applies the same approach to everyone on the team. Risk is everybody's business, especially people on the team. And finally, the last dimension is continuous process. We begin risk management at the beginning of the project, we continue doing risk assessment every week, stopping only when the project is finished, for better or for worse. Risk is continuous.

Applying these principles will greatly enhance our ability to manage and contain risk.

11.2. Risk Planning

Now let us get down to business and begin by identifying specific risk factors. Following the IEEE SPMP, we can subdivide risks into half a dozen canonical categories:

- real-world risks
- technology risks
- size risks
- people risks
- customer risks
- business risks

Real-world risks are risks that arise out of the real-world such as contract risks, management risks, external risks, and Force Majeure (aka Acts of God). Contract risks we will cover in more detail in the next chapter. Suffice it to say that anytime we sign a contract, we have significant risk exposure because of signing that contract. Management risks affect the overall health of the project. What is the financial health of the organization? Is the organization ripe for takeover? What is the cost of capital and the cost of raw materials to the organization? All of these risks are management risks that can impinge upon the project. The third type of real-world risk is external risks perhaps more precisely global risks. For example, unrest in Iran will almost always drive the price of oil up significantly. An extremely unusual but important event like Y2K can suck up all the available COBOL programmers in North America, making it impossible to populate a project team. The final real-world risks are Acts of God risks; events that we do not understand or cannot control. We tend to blame God, which is a little unfair, for risks such as hurricanes, tornadoes, tidal waves, plagues, plane crashes, and earthquakes. All of these are Acts of God risks which can have significant effect on a project. The 2011 tsunami in Japan, for example, is the perfect example of an Act of God risk.

Technology risks are risks associated with a particular technology that we are using on the project. It can involve software, hardware, or in general, new technologies that have not been tried before. New technologies can fail. The technology that we are using can be scooped or replaced by a more advanced technology. For example, in the wireless communication area, advances are regularly scooping and thus obsoleting, legacy equipment.

Risks due to size are important too. Anytime we build something that is larger than what has been built before, we have a huge risk just because the project is large. We call this the Star Trek syndrome; to boldly fail where no one has failed before. The size of the work that we are attempting to do may overstretch the technical platform that we have used in the past. Be very careful in extrapolating solutions to areas to which they have not been previously applied.

People risks are always risks that must be considered. We have problems with people leaving the project, people burning out, shrinkage of labour supply, inappropriately skilled people in the project team, and the execution of dull and boring work. A factor which most of us do not consider in project management is the relationship of trade unions with the work that we are doing. If your organization has a difficult relationship with a particular union, this will be a significant risk. The management of relationship between engineers and unions is a complicated matter and is beyond the scope of this book.

Customer risks are risks that the customer brings to the table. Is the customer financially stable? Is the customer imposing requirements changes on the project? Will the customer demand environmental changes? For example, the customer may submit request that all the documentation must be in a second language. The customer may want an external interface change. She has changed her software platform from Windows to a Linux environment and expects you to do the same. All of these are risks associated with the customer that can impinge on the project.

Business risks, like death and taxes, are always present. We may be building a product that nobody wants to use. The product may no longer fit into the strategy of our customers' vision. We may be building a product that the salesforce does not understand, and thus cannot sell. We always face the possibility of losing senior management support, which means the project funding will dry up.

To sum up, use the above general risk categories as a starting point to define the risk categories of your particular project. This leads to the construction of the Risk Breakdown Structure (RBS) as shown in Figure 11.1 — RBS for the SPENTFUEL project.

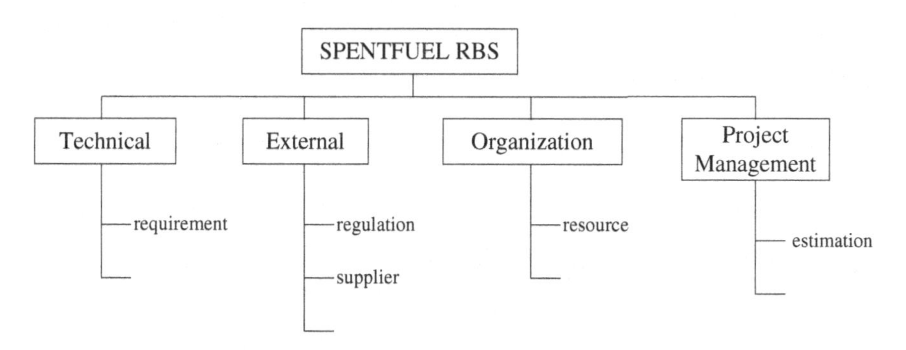

Figure 11.1. SPENTFUEL risk breakdown structure example.

The above example shows how risks in the project are categorized into technical, external, organization and project management perspectives.

Other inputs to the risk management planning process include inputs from the organization itself. The organization may be a risk-averse, like Ontario Power Generation (OPG). It may have policy statements that are generic for the entire organization. The exact position on risks that company takes may be inferred from previous actions of the company. For example, innovative IT companies such as Apple would take much more risky actions than an established organization like OPG. The organization may have predefined risk categories, which means we have to, of course, use them. There may be standard templates. It is a good idea to indicate the authority levels for risk decision-making in the organization as we will need to add this into the plan. As an example of generic risk in software engineering, Table 11.1 shows a list that Barry Boehm defined for Boeing as the top 10 generic software risks [2].

Table 11.1. Boehm's top ten software engineering risks.

Risk Rank	Risk Management Techniques
1. Personnel shortfalls	Staffing with top talent, job matching team building, morale building, cross-training, prescheduling key people
2. Unrealistic schedules and budgets	Detailed, multisourced cost-schedule estimation, use of design and FP metrics, software reuse; requirements engineering
3. Developing the wrong software functions	Organizational analysis; mission analysis; user surveys; prototyping; early user's manuals
4. Developing the wrong user interface	Task analysis; prototyping; scenarios; user characterization
5. Goldplating	Requirements scrubbing; prototyping; cost-benefit analysis
6. Changing requirements	High change threshold; information hiding; incremental development
7. Shortfalls in externally supplied components	Benchmarking; inspections; reference checking; compatibility analysis
8. Shortfalls in externally performed tasks	Reference checking; pre-audit awards; award-fee; team building; competitive design
9. Real-time performance shortfalls	Simulation; benchmarking; modeling; prototyping; instrumentation; tuning
10. Straining technical capabilities	Technical analysis; cost-benefit analysis; prototyping; reference checking

Table 11.2 shows another list of generic software engineering weaknesses:

Table 11.2. Generic SW weaknesses [6].

1.	Schedules set before requirements definition
2.	Excessive schedule pressure
3.	Major requirements change after sign-off
4.	Inadequate project management skills
5.	Inadequate pretest defect removal procedures
6.	Inadequate software process
7.	Inadequate office space, environment
8.	Inadequate training
9.	Inadequate support for reuse, code and design
10.	Inadequate organizational and specialist support
11.	Too much emphasis on partial solution
12.	Too new technologies

Here is the template for the Risk Management Plan from Charette [7]:

I Introduction
1. Scope and purpose
2. Overview
   (a) Objectives
   (b) Risk aversion priorities
3. Organization
   (a) Management
   (b) Responsibilities
   (c) Job descriptions
4. Aversion program description
   (a) Schedule
   (b) Major Milestones and reviews
   (c) Budget

II Risk Analysis
  1. Identification
    (a) Survey of risks
    (b) Sources of risk
    (c) Risk Taxonomy
  2. Risk Estimation
    (a) Estimated probability of risk
    (b) Estimated consequence of risk
    (c) Evaluation of risk referents
    (d) Evaluation results
  3. Evaluation
    (a) Evaluation methods to be used
    (b) Evaluation method assumptions and limitations
    (c) Evaluation risk referents
    (d) Evaluation results

III Risk Management
  1. Recommendations
  2. Risk aversion options
  3. Risk aversion recommendations
  4. Risk monitoring procedures

IV Appendices
  1. Risk estimate of the solution
  2. Risk abatement plan

11.2.1. Risk identification

Now it is time to bell the cat; that is define the risk precisely. In risk identification, there are two steps. Step one is to capture the statement of the risk. Step two is to establish the context of the risk. The statement of risk consists of two parts. Part one is the condition. The condition is a single phrase that describes the key circumstances, situations, cause of concern, doubt, anxiety, or uncertainty. The second part, the consequence, is a single sentence describing the key outcomes, normally negative, of the current situation. For example, suppose we were talking about the Graphical User Interface used for the SPENTFUEL project. The condition of our risk might be: the GUI must be coded using X Windows, and we do not have experience in X Windows. The consequence of the statement of the risk is: the GUI code may not be completed in time and may be inefficient.

The second step in risk identification is capturing the context of the risk. This includes recording additional information regarding the related circumstances, events, and interrelationships which may affect the risk. The intent is to provide extra information that other folks can understand what it was that caused us to worry about the risk, particularly after time has passed. Remember some of the risks identified today are fleeting. They will not be risks five years from now, or the context will be completely unintelligible to somebody trying to understand what you are worried about. For example, anybody that has been in the software business for 20 years knows what a GUI is. People just entering the industry likely do not. So we would start off with capturing our context: something like "the graphical user interface is an important part of the software system and we do not have anyone trained in X Windows. X Windows is an extremely complicated window system developed at MIT. We have all been studying the language, but it is complex, and only one person on the team has any graphics experience and that is with a primitive Windows system on a PC." There is the context of the risk so that a person can understand it, several years down the road.

Risks can be identified by some of the following techniques:

- Documentation review
- Facilitation techniques
    - Brainstorming — follow a round robin token passing process to solicit ideas from participants, all inputs count and no idea is a bad idea
    - Delphi — method using SMEs
    - Expert interview — one-on-one discussion with experts
    - Root cause analysis — examine the root cause of problems
- SWOT — examine from strength, weakness, opportunity, and threat perspective

- Checklist
- Assumption analysis — every assumption and constraint is a risk
- Diagramming — fish bone, flow chart, etc.

There has been a suggestion that risk identification be done in three iterations. A risk management team (e.g., team leaders, key project personnel) will initially participate in the first iteration, followed by the entire project team and stakeholders in the second iteration, and finally, persons not involved in the project, will provide an unbiased analysis in the third iteration. In practice, only iteration one or iteration two is commonly used in industry.

The major output of the risk identification phase is the Risk Registry. It is a list of all the identified risks. It will list the tracking metrics and the triggers associated with the risk. For example, what is the parameter that will indicate a risk is coming close to eventuate? And what is the trigger value that will force us to invoke a mitigation option? Included is a list of potential responses, root causes of those risks, risk owners, and updated risk categories. In essence, anything associated with risk identification will be recorded into the risk registry. The Risk Registry, of course, will be available to other planning areas and processes.

11.2.2. Risk qualitative analysis

The next phase of our analysis, now that we have identified our risks, is to rank risks relative to each other. We ask the question, what is the most important risk of all to be considered? We may either assign numeric values to the risks or qualitative values. Each has its own place in doing risk analysis. First, let us consider risk qualitative analysis. It focuses on ranking the risks in non-numeric terms. We want to identify what a high priority risk is. Qualitative risk analysis is rapid, cheap, and foundational. It points in the direction of the quantitative in certain areas as we will see later. Risk has two measurable attributes: the likelihood of the risk and the consequence of the risk. The combination of these two, we refer to as the severity. The severity is equal to the likelihood times the impact. So if we had a risk, for example, that had a high likelihood of occurring, and had a devastating impact on the project, we would rank that as more serious than a risk that had a low likelihood and a low impact. There are several ways we can do qualitative risk analysis. The first way is to use a binary approach. We take a look at our likelihood — impact characterization of the risk and we assign a binary value to it. The likelihood of the risk occurring can be either high or low. The consequence of the risk can be high or low. The combination of these gives us four categories. That is,

	Likelihood	Consequence	Rank
1.	High	High	high risk
2.	High	Low	moderately high risk
3.	Low	High	moderately low risk
4.	Low	Low	low risk

Thus using the binary approach to risk ranking, we rank all our risks in terms of one to four with one being the most severe.

Other approaches are possible too. We can use a ternary approach where the rankings now are high, medium or low. The corresponding ranking table gives us a ranking from one to nine.

	Likelihood	Consequence	Rank
1.	High	High	very high risk
2.	High	Medium	high risk
3.	Medium	High	somewhat high risk
4.	High	Low	high medium risk
5.	Medium	Medium	medium risk
6.	Low	High	high low risk
7.	Medium	Low	low medium risk
8.	Low	Medium	low risk
9.	Low	Low	very low risk

Finally, we can use the Air Force quaternary categorization system. The likelihood is defined to be frequent, probable, improbable, or almost impossible. The consequences are similarly ranked in four: catastrophic, critical, marginal, or negligible. This gives the 16 possible boxes into which to place the risks. We leave it as an exercise for the reader to fill this out.

The fourth way of assigning risk is to generate what is called the Risk Product Number (RPN). Instead of using two variables, it uses three. It uses the occurrence and severity, which are renames of our previous likelihood and consequence variables, plus a third variable called detection. It then requires a ranking from 1 to 10 of each of these three variables and produces a product of the three of them. So for example, if you had an occurrence of 1, a severity of 1 and a detection of 1, the RPN would be 1. If you had an extremely risky product, with the ranking of 10 for occurrence, 10 for severity and 10 for detection, the RPN would be 1000: yikes, head for the hills!

For the Occurrence ranking, you could rank them as follows:

- Remote failure rate 1
- Low failure rate 2–3
- Moderate failure rate 4–6
- High failure rate 7–9
- Almost certain failure 10

For the Severity ranking, you could rank them as follows:

- Insignificant 1
- Low severity 2–3
- Moderate severity 4–6
- High severity 7–9
- Very high severity 10

For the Detection ranking, you could rank them as follows:

- Very high 1
- High 2–3
- Moderate 4–5

- Low 6–7
- Very low 8–9
- Unlikely 10

Any of the four ways allow us to rank the risks and to place them in a priority list. This is called Comparative Risk Ranking (CRR). An illustration is given in Figure 11.2.

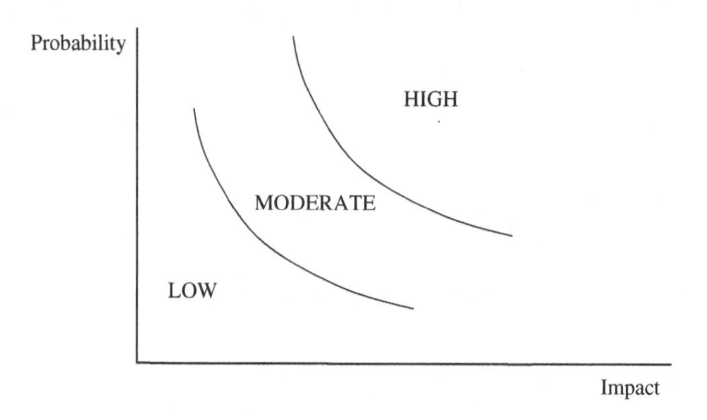

Figure 11.2. Probability impact assessment matrix.

As a rule of thumb, we can actively track at most 10 risks. Whatever the ranking, we take the ten highest risks and these will be the ones on which we will place most of our efforts.

11.2.3. Risk quantitative analysis

Based on inputs from the time management plan and cost management plan, and life cycle costing (LCC) figures (such as return on sales (ROS), return on asset (ROA), net operating profit after tax, etc.), risks can be analyzed in a quantitative manner. The formula for expected monetary value (EMV) analysis is given below.

$$\text{Expected value} = \text{impact} \times \text{probability of occurrence}$$

For example, if within one year of paving a driveway, the driveway cracks with a width of over 2 inches, the paving company will be fined \$1000, and the probability of this occurrence is 10%, then EMV = \$1000 × 10% = \$100.

Sensitivity analysis helps determine which risks have the most impact. Decision tree analysis helps determine the probability of occurrence. Monte Carlo simulation blends the impact and probability together with some randomness in the calculation of expected value.

Performance Evaluation and Review Technique (PERT) is a triangular distribution type estimate using the three-point input for optimistic (o), most likely (m), and pessimistic (p) views.

$$\text{PERT result} = \frac{o + 4 \times m + p}{6}$$

$$\text{Standard deviation, sigma} = \frac{p - o}{6}$$

$$\text{Variance} = \text{square of sigma}$$

Note that this calculation forces a normal distribution with 68% of data within the "o" and "p" range.

Consider an estimate to finish a project with t_o = 9 months, t_m = 10 months, t_p = 1 year; using PERT calculation:

$$t_E = \frac{9 + 4 \times 10 + 12}{6} = 10.2 \text{ months}$$

$$\text{Sigma} = \frac{12 - 9}{6} = 0.5 \text{ month}$$

There will be 68% probability of finishing the project within 9.7 to 10.7 months (i.e. 10.2 ± 0.5).

There will be 95% probability of finishing the project within 9.2 to 11.2 months.

There will be 99% probability of finishing the project within 8.7 to 11.7 months.

11.2.4. Risk response planning

This is the process to develop options and determine actions to reduce threats (negative risks) and pursue opportunities (positive risks).

The strategies for threats are:

- Avoidance (aka prevention in PRINCE)
  For example, buy a house on high ground to avoid basement flooding.
- Transference (deflection)
  For example, buy insurance that covers basement flooding.
- Mitigation (aka reduction in PRINCE)
  The strategy will attempt to minimize the probability or impact or both. For example, keep expensive furniture and carpet away from the basement to minimize the lost (impact) in case of flooding. Install drain pipes around the house to minimize the probability of water leaking into the basement.

The strategies for opportunities are:

- Exploit
  Pursue the opportunity so that it comes true.
- Share
  Allocate ownership to third party to capture the opportunity. For example, form partnership (joint venture) with an international company to market your product while your company is still localized.
- Enhance
  Maximize the probability or impact, or both.

Other strategies for dealing with threats or opportunities are:

- Passive acceptance
  Wait till it happens and still do nothing about it. For example, you accept the risk that buying a house in California may suffer from non-recoverable damage in case of a major earthquake.
- Active acceptance
  Same example as above, yet you have a shelter containing food and water.

11.3. Risk Executing

In this phase, corrective actions are defined which, if the risk is eventuating, could be executed to mitigate the consequence of the risk.

- Contingency or prevention action refers to planned corrective actions to negative risk events
- Workaround refers to unplanned corrective actions to negative risk events

Fall backs, unlike corrective actions, mean changing the project scope and developing alternate options; in other words, changing the PMP.

11.4. Risk Monitoring and Controlling

Risk monitoring and controlling can be achieved by some of the following:

- Status meeting

  Help identification of additional risks during project execution and check progress in tackling existing risks. For example, in a one-hour weekly status meeting with the entire project team, allocate 10 minutes to go through the top 5 risks.
- Risk re-assessment

  This is needed from time to time as the probability of occurrence and impact may change.
- Risk audit

  Are risks being identified, qualified, and quantified? Is the risk management plan being followed? Are corrective actions and fall backs being executed with progress being tracked just like any other schedule activities?
- Variance and trend analysis

  Exceptions to acceptable values need a round of risk re-assessment.

- Contingency/Reserve analysis
  Check to see whether the typical 10% project reserve fund has been used up. In exceptional cases, we may have to dip into the organizational reserve fund.

Generally speaking, when a negative risk materializes, there is a problem. When a positive risk materializes, there is a windfall.

11.5. Risk Closing

Use the Risk Database (Issue Log, or Risk Registry) as the basis to conduct a lessons learned exercise.

11.6. PERT Exercise

Fill in the following table based on the PERT estimates given for a software development project:

	Optimistic	Most likely	Pessimistic	PERT result	Standard deviation
Requirement	3 weeks	4 weeks	6weeks		
Design	5 weeks	6 weeks	10 weeks		
Coding and unit testing	5 weeks	6 weeks	8 weeks		
Integration test	2 weeks	3 weeks	4 weeks		

- o What is the optimistic amount of time for the project?
- o What is the most likely amount of time for the project?
- o What is the pessimistic amount of time for the project?
- o What is the PERT calculated amount of time for the project?
- o What is the probability of finishing the project within the PERT calculated amount of time?
- o What is the amount of time for the project to finish with 84% confidence level?
- o Which phase of development is the most risky? Why?

11.7. Multiple Choice Questions

1. Which of the following is not a tool to identify risk?
   (a) Delphi
   (b) SWOT
   (c) WBS
   (d) Brainstorming

2. What does passive acceptance of risk mean?
   (a) Mitigate the risk by minimizing its probability of occurrence
   (b) Mitigate the risk by minimizing its impact
   (c) Have a contingency plan ready for execution
   (d) Take no action

3. Which project management process does Perform Quantitative Risk Analysis under Project Risk Management belong to?
   (a) Initiation
   (b) Planning
   (c) Execution
   (d) Control

4. Perform Quantitative Risk Analysis includes:
   (a) Enumerating sources of internal and external events
   (b) Identifying potential events and impact
   (c) Evaluating probability and impact
   (d) Developing contingency plans and resources

5. In which phase of the project do you have the greatest influence on project risk?
   (a) Initiation
   (b) Planning
   (c) Execution
   (d) Termination

6. A contractor's deliverable has been delayed by 30 days. The process of determining how this event will affect the project schedule is called risk:
   (a) Identification
   (b) Mitigation
   (c) Simulation
   (d) Assessment

7. An individual's willingness to take a risk can be determined by:
   (a) Decision tree modeling
   (b) Monte Carlo method
   (c) Sensitivity analysis
   (d) Utility function

8. Your old house was right next to a river. You had bad experience of flooding in the basement whenever it rained heavily. When you decide to purchase a new house, you have made up your mind to only consider houses on high grounds. You have reduced the threats of the risk of flooding by:
   (a) Avoidance
   (b) Transference
   (c) Minimizing the impact
   (d) Active acceptance

9. Plan Risk Responses is intended to:
   (a) Create steps to identify project risks
   (b) Formulate strategies for dealing with adverse events
   (c) Construct a list of previous project risks
   (d) Develop measurements to quantify project risks

10. According to the PERT calculation, what is the standard deviation for the 3-point estimate of the following activity?
    Optimistic: 4 days; most likely: 5 days; pessimistic: 10 days
    (a) 1 day
    (b) 5 days

(c) 5.67 days
(d) 6 days

11. According to the PERT calculation, what is the probability of finishing the following 3-point estimate activity within 7.67 days?
Optimistic: 4 days; most likely: 5 days; pessimistic: 10 days
(a) 34%
(b) 50%
(c) 68%
(d) 95%

12. According to the PERT calculation, what is the probability of finishing the following 3-point estimate activity between 4.67 days and 6.67 days?
Optimistic: 4 days; most likely: 5 days; pessimistic: 10 days
(a) 34%
(b) 50%
(c) 68%
(d) 95%

13. Your company buys insurance in excess of minimum government requirements for employees working outdoors. The company reduces threats of the risk of outdoor injuries by:
(a) Avoidance
(b) Mitigation
(c) Deflection
(d) Minimizing probability of occurrence

14. The highest risk impact occurs during which of the following project life cycle phases?
(a) Initiation and planning
(b) Planning and execution
(c) Execution and closure
(d) Initiation and closure

15. If the odds of completing activities 1, 2 and 3 on time are 50%, 50% and 50%, what is the chance of starting activity 4 on day 6?
    (a) 10%
    (b) 13%
    (c) 40%
    (d) 50%

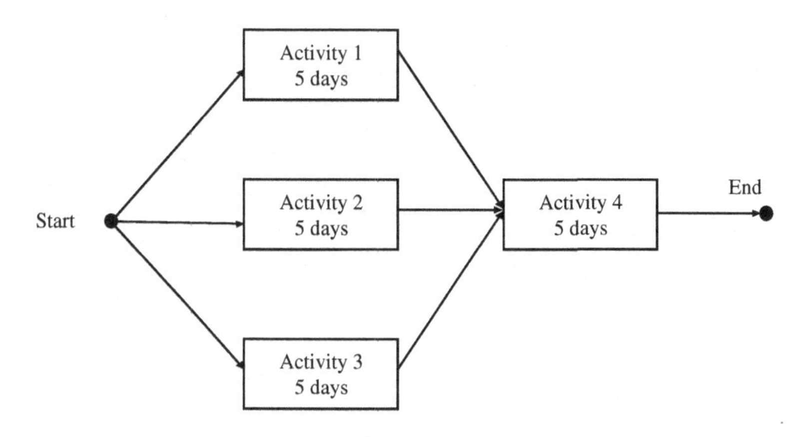

References

1. Szu, S. *The Art of War*. Simon and Brown, 2012.
2. Pressman, R. *Software Engineering: A Practitioner's Approach*, 7th Edn. McGraw-Hill, 2010.
3. Project Management Institute. *A Guide to Project Management Body of Knowledge* (PMBOK), 5th Edn. PMI, 2013, p. 559.
4. PRINCE2 Electronic On-Line Manual. Central Computer and Telecommunications Agency (CCTA).
5. Software Engineering Institute. *Principles of Continuous Risk Management*. Pittsburgh, PA: SEI, 1995.
6. Jones, C. *Estimating Software Costs*, 2nd Edn. McGraw-Hill, 2007.
7. Charette, R. *Software Engineering Risk Analysis and Management*. McGraw-Hill, 1989.

CHAPTER TWELVE

Procurement Management

Project procurement management is an umbrella term that includes all the processes needed to purchase or acquire services, products, and/or results from outside the team. It refers to the organization of work that others will do for us. It can range all the way from something quite simple, such as the purchase of a box of paper clips, to outsourcing a large piece of software, like the database module of SPENTFUEL. Other names for procurement management are: contract management, contracting in general, tendering, etc.

There is an important implication of procurement management that we have two entities; a buyer and a seller, bound together by a contract. The buyer is buying the services of the seller, with the contract specifying the obligations of both sides. Note that our organization can be either a buyer or a seller, depending on the roles that we are playing. Various other terms are in use, such as owner/contractor in the construction industry and offerer/offeree in the legal industry. We will use buyer and seller to indicate that somebody wants work being done (the buyer) and the seller, who will do the work for the buyer at a set price. Procurement management is the process of making sure that this gets done efficiently, on time, on budget and on quality.

There are four major processes in procurement management.

12.1 Procurement planning — the process of documenting the decisions that led us to do a procurement in the first place, specifying that particular contract we have in mind and identifying potential sellers.

12.2 Procurement executing — the process of obtaining responses from various sellers, picking a seller that can best satisfy our needs and awarding a contract to get the work done.

12.3 Procurement monitoring and controlling — this is the process of managing the execution of procurement, awarding and administering the contract, making sure we have appropriate contract performance, and making corrections and changes as needed.

12.4 Procurement closing — the final process of wrapping up the procurement process, documenting our lessons learned and paying off the contractor.

We also cover outsourcing and offshoring in 12.5, as well as contract law in 12.6.

12.1. Procurement Planning

The purpose of the procurement planning phase is threefold: to generate the procurement management plan, to produce the procurement Contract Statement of Work (CSOW — a narrative description of the product or service to be provided by the seller to buyer), and to produce the procurement documents which include the source selection criteria. The buyer determines what, when, and how to procure. A make/buy analysis has to be done by looking at the direct and indirect costs. The project purchasing decisions are documented, the approach is specified and the potential sellers are identified. Expert inputs can be solicited through other units within the organization, consultants, professional associations, industrial groups, market research and meetings, etc. Inputs into the process include the usual suspects: scope baseline, requirements documentation, teaming agreements, the risk registry, activity cost and time estimates, and any environmental factors that arise from the organization.

Teaming agreements require a slight expansion. These are contractual agreements between two or more entities to form a partnership or joint venture to do something such as part of the contract. When they are in place, they define the roles of the buyer and the seller, and so on. The US Defense Contract Audit Agency defines teaming agreements to be "an arrangement between two or more companies either as a partnership or joint venture, to perform on a specific contract. The team itself may be designated to act as a prime contractor; or one of the team members may be designated to act as a prime contractor, and the other members designated to act as subcontractors."

We assume that we have decided as a team and an organization to contract out a piece of the work. As a concrete example, let us assume we are going to issue a contract for the development of the database module for SPENTFUEL. The first thing we have to decide is the type of contract that best meets our needs.

Generally speaking, there are two contracting methods: competitive (sellers have to compete for the contract) and non-competitive (the same seller always gets the contract).

- Competitive contracts

  Here are some common terms used in structuring competitive contracts.

  - IFB — Invitation for Bid

    The buyer is going to award the contract to the seller with the lowest cost.

  - RFP — Request for Proposal

    Internationally known as RFT — Request for Tender, the buyer is going to award the contract to the seller with the best value (e.g., cost, quality, time to deliver, support cost, reputation, etc.)

  - Reverse auction

    Online electronic bid with open information.

Note that for IFB and RFP (RFT), the bids and proposals from the sellers are treated with confidence; unlike an online e-bid with open information. In addition, bids and proposals from the sellers to IFB and RFP (RFT) are considered as offers to enter contracts. We will have more to say on this important point in Section 12.6.

The Letter of Intent (LOI) and Memorandum of Understanding (MOU) are instruments to enter into a contract. Once a LOI or MOU is signed, good faith is mandated from then on for both sides to enter a contract. Request for Quotation (RFQ) and Request for Information (RFI) are only requests from the buyer to solicit seller information with no obligation to enter a contract.

- Non-Competitive Contracts

  The buyer will simply pick the same seller for a certain product or service.

  - Single source

    The same seller is approached to provide a certain product or service. For example, some big organizations would have completed their competitive contract assessment up front and decided on one or a few sellers for PC software; or one or a few travel agencies to book corporate business trips for employees.

  - Sole source

    Only one seller can provide certain product or legal service because of a monopoly (e.g., holding a patent).

Figure 12.1 illustrates various contract pricing categories and types within each category.

There are three broad categories for contract pricing: fixed price, cost reimbursement, and time and material.

- Fixed Price

  This category of contract pricing is for well-defined products. It is low risk for the buyer (since the price is fixed) and high risk for the seller (in case of cost overrun). However, if the work is unfamiliar to both the buyer and seller, fixed price contracts impose high risks to both.

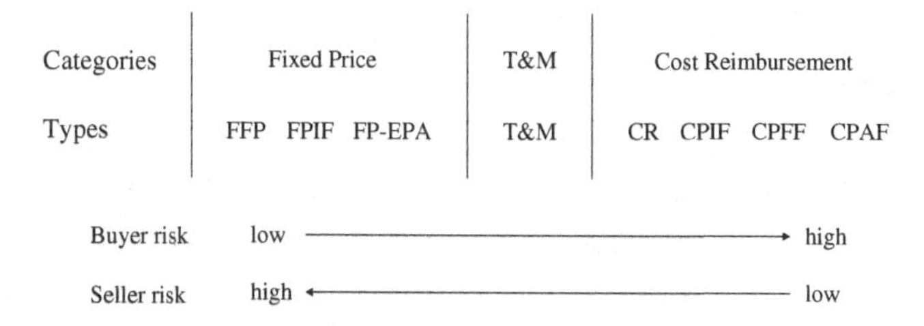

Figure 12.1. Contract pricing categories and types.

- FFP (Firm Fixed Price)

  An example is a purchase order ($10,000 for a piece of test equipment).

- FPIF (Fixed Price with Incentive Fee)

  This is typical for contract with a long performance period and of high value. The idea is to encourage the seller to keep costs as low as possible, and to share any savings at some agreed percentage with the seller. (The seller will receive the fixed price plus some percentage of the savings, as bonus, for example.) On the other hand, in case of cost overrun, the additional costs will always be the responsibility of the seller.

  For example, suppose you have been contracted to implement a project. The estimated cost of the project is $1,000,000. The contract agreement states that any cost savings will be split between you (20%) and the buyer (80%). If, at the end of the project, the cost for the project is $800,000, how much will be awarded to you?

- FP-EPA (Fixed Price — Economic Price Adjustment)

  The idea is to relate the final price to some reliable financial index (e.g., inflation) for a contract with a long performance period. Other time sensitive indices include the price of oil, the value of the Canadian dollar to the euro, or American dollar, the price of gold, the price of silver, and so on.

- Cost Reimbursement

  This category of contract pricing is high risk for the buyer and low risk for the seller.

- CR (Cost Reimbursement)
  The name is self-explanatory.
- CPIF (Cost Plus Incentive Fee)
  This is typical for a contract with a long performance period and substantial hardware and test requirements. The incentive is that any savings that occur will be split at a specified ratio, normally 50-50.
- CPFF (Cost Plus Fixed Fee)
  This is typical for research and development (R&D) projects.
- CPAF (Cost Plus Award Fee)
  This contract type specifies the majority of fee earned to be based on satisfaction of broad subjective performance criteria. The award is not subject to appeal.

- Time and Material
  This is a hybrid type of arrangement combining both fixed price and cost reimbursement types. For a well-established industry like car repairing, T&M contracts are appropriate. For unfamiliar products and services, the buyer should carefully audit the hours as this category of contract pricing tends to reward inefficiency. The seller receives a higher return just because longer time is spent on the job with therefore more material being used.

The last piece of procurement planning is the construction of the procurement management plan. This will include the type of contract we have chosen, management of risk issues, whether we are going to use independent estimates or whether we are going to trust the seller to provide the estimates, the use of standard templates, and details associated with the management of the contract. The procurement management plan will identify the need for performance bonds. It will establish standards that the seller has to use, for example, earned value management. It will also contain the CSOW. There has to be sufficient detail for the seller to be able to accurately bid on the work and give precise cost and time estimates. The procurement management plan will also include the schedule for the solicitation of proposals from prospective sellers.

Typically, we would issue two types of documents. One type is legally non-binding such as the RFI or RFQ. When we do not know whether or not any seller can perform the work, we need to test the water so to speak, and see if there are potential bidders out there we can depend on and roughly how much do we expect to spend. The other type is legally binding such as the IFB or RFP/RFT. This is the document that we are sending out and expecting the seller to respond to it. We are going to specify the source selection criteria. If someone meets the source selection criteria, we are mandated to award the contract. Specifying the criteria for rating the proposals, therefore, is very important.

Criteria may be either mandatory or desirable. Any criterion that is mandatory must be met. If a single criterion is not met, that seller's proposal cannot be accepted. Desirable means you might achieve extra points for adding features that are deemed useful, but not mandatory to the buyer. Other selection criteria that we might invoke include:

- understanding the need
- the overall life cycle cost
- technical capability
- management approach
- technical approach
- financial capability
- production capacity and interest
- business size and type
- references
- intellectual property rights issues
- proprietary rights

The most common way to compare different responses is to use an arithmetic scoring model. We would assign a weight to each of the preceding attributes. We would then define a metric for each one of those attributes, likely a Likert scale from 0 to 5. On receipt of the responses from the sellers, we would measure each metric for each of the attributes and sum them up, modified by the weighting factors. The vendor scoring the largest number would be awarded the contract.

Now we are in a position to solicit sellers to do the work.

12.2. Procurement Executing

Now that we have planned the work, we work the plan. The first thing is to send out the request to interested parties. We would obtain some of these names from previous work within the organization. Other sources would include checking the Internet, checking professional societies, checking friends of the organization, etc. It is useful to have a bidders' conference (aka pre-bid conference, contractor/vendor conference) before the actual evaluation of the tenders. Points of clarification can be made that either side may not have thought of. For example, when is the work expected to be completed? If it is indeed part of the project, it is likely there is a fixed date by which the work has to be done. If that is not included in the request, then we need to clarify it at the bidding conference. It is vitally important that any information not included or not clear in the request, but clarified or modified at the bidders conference, is written down and made available to any responding seller who might not be able to make it to the bidders' conference.

In reviewing offers, the buyer may engage its procurement organization to prepare its own independent estimate (aka "should-cost" estimate) for comparison purpose. In choosing potential sellers, the weighted attributes evaluation is commonly used. The attributes of interest are first identified (e.g., technical feature of the product, size of the company, price of the product) and the weights for the attributes are given. The buyer must inform the sellers whether evaluation is weighted, but there is no need to disclose the factors (e.g., price gets the highest weight). Some standards of evaluation are:

- absolute

  Disqualify seller if a criterion is not met within a maximum or minimum range.

- minimum

  A screening system to disqualify a seller if a certain minimum criterion is not met.
- relative

  If one is not familiar with any maximum or minimum standards for evaluation, just gather the data and compare the relative values from various sellers later. For example, if you are buying a private jet, do you know the speed requirement for the plane?

We will select the winner according to the selection criteria that we have defined in the procurement management plan. We must do this. Failure to do this means that we are not following our own rules and this would be interpreted as illegal in the courts. We might want to make changes to the contract. For example, the seller may be able to do it a month or two faster if we spend a bit more money. If that is acceptable to us, we can make little changes to the contract as long as the changes do not go to the heart of the contract. Changes that go to the heart of the contract are not permissible.

12.3. Procurement Monitoring and Controlling

In this phase, we award and administer the contract. The contract will contain the major milestones in the running of the project. For example, regular performance reviews must occur. If earned value is mandated in the contract, those numbers have to be reported on a periodic basis. The results of inspections and audits will be made available to the buyer, so that there is assurance of work being done appropriately. Normally, payments are structured so that just enough money is released to permit the seller to proceed to the next phase, and no further. If there are changes that have to be made in the contract, then the procedures specified in the contract must be followed. It is permissible to change a contract if both sides agree, but it is not permissible for one side to unilaterally make major changes as we will see a bit later. It is the responsibility of the person in charge of the project to keep on top of the seller to make sure the work is being done in an appropriate manner. Often the pointy haired boss thinks that just because we have subcontracted out the work to the seller, no further effort is needed on the part of the project team. Nothing could be further from the truth. The fact that we have outsourced a piece of work does not relieve the PM of the obligation to make sure the work is done appropriately. A nice way to view this is that the PM is now in charge of the team that just happens to be paid by another organization, not his own. Outsourcing projects that work well invariably include a good working relationship between the seller's PM and the buyer's PM — the two of them maintaining close contacts, meeting at least weekly to discuss progress, or lack thereof. If they hit an impasse that they cannot resolve, then each side will escalate its concern to his superior, and the two superiors will then decide which way the argument should go.

Although both the buyer and seller are obligated to comply with the terms and conditions of the contract and to resolve claims and disputes to the best of their abilities, there may come a time that unsatisfactory work leads to disputes, or contested changes, resulting in contract termination. Dispute resolution may go through the following channels in steps:

- Negotiation
  Also known as compromise, a third party will be involved.
- Mediation
  This will involve a mediator.
- Arbitration
  This will involve an arbitrator. Arbitration is *fait accompli,* or the topic of dispute reaches final decision and cannot be further escalated. It is a one-time shot with no appeal, unless both sides agree to further litigate. For international contracts, arbitration is preferred as most countries participated in an international convention that allows enforcement of the international arbitration ruling.

- Litigation
  Litigation has many appeal opportunities and is a lengthy and costly process. In addition, the ruling in one country is usually not applicable to another country.

One final remark is that in today's world with lots of acquisition, the assignee (acquirer) is responsible for the contracts of the assigner (party being acquired).

12.4. Procurement Closing

The procurement will be closed in one of three ways: premature, scheduled, or overscheduled. In all three cases, the reasons and the indicators for closing must be documented. Contract closeout involves product verification, procurement audit, and administrative closeout. We have to have official verification that all the work that has been done to date has been done satisfactorily. In the case of the project being closed on a scheduled basis, once that assurance has been given, we may close the work and pay off the contractor or seller. The contract is then completed and settled, and open items are resolved. The person (not necessarily the PM) responsible for contract administration provides formal notice of contract completion. Then we must conduct a lessons learned on what did we do right. What could we do better? What should we avoid next time? All of these questions need to be answered in an effective way so that we can learn from the work that we have done to improve the execution of the next procurement.

12.5. Outsourcing and Offshoring

Outsourcing refers to procurement of products or services from a seller. The seller can be an external vendor or a previous in-house sector or division that has been sold to another company or spun off as a separate company. In recent years, companies in various fields, especially IT [1], have jumped on the bandwagon whereby outsourcing is further refined with offshoring in two ways — Intra-Firm Offshoring [2] to an offshore affiliated entity that the company itself owns, or Third-Party Offshoring to an offshore unaffiliated company.

12.5.1. Outsourcing, insourcing, offshoring, and near-shoring

Build or buy? That is the issue with outsourcing. When a company needs to add a new service, it has to decide whether it can build it in-house or whether it should get somebody else outside the organization to build it. Suppose we needed to implement a help-desk; should we implement it internally or externally? What about a backup service? What about a new technology that we have never experienced before? Outsourcing in IT particularly began with what we now call the Kodak effect. In 1989, Kodak signed five- and ten-year outsourcing agreements valued almost $1 billion with three companies. It was running into significant financial difficulties at the time and its IT department was not effective, so that year it outsourced all of its IT services. The CIO of the company became very famous and got her picture on a lot of trade journals. But within five years of signing the outsourcing agreements, Kodak laid off 10,000 people. By 2010, the company was virtually bankrupt. One could blame it on the outsourcing initiative. It certainly did not help the company.

When should you outsource? Some obvious answers are to reduce headcount, to implement a service that is common with other competitors that can do it better and more cheaply, when you cannot afford the time or money for implementing a new technology, when you cannot keep good staff, when you cannot get good staff, when the project is small and well defined.

The opposite of outsourcing is insourcing. You would do that when the function you need to implement is mission-critical. If something is a core competency of the organization, it should be insourced. If you cannot identify any cost savings, then you should insource. If it is not clear that the vendors are better than you are at building the product, do it in-house, and especially if the vendor cannot be controlled, then you should never outsource. The third choice is to selectively outsource. When you have a shortness of time or staff for a small project, it may make sense to outsource for that particular case. If you can make the ROI, then outsourcing may be a good thing to do.

Recently we have made a distinction between offshoring and near-shoring; the former being a contact with the country beyond the physical boundaries of the continent, the latter being a country physically contiguous with the main organization.

12.5.2. Outsourcing and offshoring initiating

Just like running any project, outsourcing/offshoring naturally follow the life cycle of initiating, planning, executing, monitoring and controlling, and closing. Each process group needs additional consideration on top of the standards and procedures of running a regular project. Think first! Clear objectives have to be set [3]. Potential for performance improvement and business implication must be assessed. The project needs to be formally sanctioned by a sponsor.

12.5.3. Outsourcing and offshoring planning

In order to plan for and subsequently manage and control outsourcing/offshoring, the procurement process needs to be established. An agreement needs to be packaged to define the service requirements, covering both the transition and subsequent developments. For intra-firm outsourcing/offshoring, a Service Level Agreement (SLA) will suffice; for third-party outsourcing/offshoring, there should be a formal contract. Specific legal agreement should also cover the ownership of artifacts or assets developed by the offshore entity, especially for assets developed by the third-party company. Who owns the assets and how are patents to be treated? In any event, a centralized intellectual property repository has to be established. How about the confidentiality of data and information in general?

In addition, the Human Resource Department has to be engaged as it has to prepare for hiring in the intra-firm offshoring entity and firing (i.e., separation) in the local entity. Terms and conditions need to be agreed for knowledge and skills transition to the offshore entity prior to separation. For example, the severance package is payable upon successful transition of knowledge and skills to the offshore/outsourcing entity.

You cannot outsource your responsibility [4]. For the local entity, get your house in order! It needs to take on new management responsibilities that do not already exist in a typical firm [2]. Start with exercising due diligence and prepare a Transition Project Management Plan (TPMP). The TPMP (just like a PMP) with extension to cover the transition for outsourcing/offshoring should minimally contain the following:

- Scope

  For the transition project, what criteria need to be satisfied to declare successful completion of the transition? How are changes in the scope of transition to be accepted or rejected? For subsequent development, how are scope changes going to be handled? Does the local entity, offshore entity or both entities have the say on acceptance or rejection of scope changes, and hence, the impacts on other artifacts?
- Time

  All transition activities such as preparing the TPMP, knowledge transition and training, pair programming, etc. are captured in a detailed schedule and baselined.
- Cost

  Extra cost for travel must be allocated. Should there be hardware or software transfer, the charging formulae need to be defined. For example, if hardware, test equipment and software licenses are to be transferred to the offshore entity, how do the existing license agreements with vendors accommodate that? How is the offshore entity going to pay the local entity upon receiving the hardware, equipment or software?

- Quality

  It is most likely that quality is going to suffer as the offshore entity is picking up the knowledge for project development for the very first time. How should the quality goals and hence, the cost of development (reflecting higher cost of quality, CoQ, due to more testing and bug fixing), be adjusted? The cost estimation for the offshore entity may follow an S-curve with higher rise at the tail, reflecting cumulative effort from a rear-loaded Rayleigh Model [5].
- Human Resource

  Training for cultural diversity or even "do's and don'ts" for outsourcing/offshoring should be organized. Staff in the local entity will develop new skills (such as product and service transitioning skills) and for those being kept, new roles and responsibilities need to be defined to maintain specialty technical functions in-house. For staff being displaced, training should be provided to ease career transition.
- Communication

  Identify all team members on both sides. List the focal points of contact and the point of escalation to minimize the overhead of communication. Clearly define the governance structure (i.e., escalation process in case anything goes wrong during the transition) and focus on relationship management. Which entity is going to run the status meeting on the transition project? Who publishes status reports on the transition and provides updates to the sponsor and stakeholders? Who drives subsequent development? Expect difficulties in managing across different time zones and cultures [1].
- Risk

  Of course, there are additional risks associated with the transition. Identify, mitigate and monitor these risks. Note that not all risks are controllable, as there are internal and external risks of the offshore entity [6].
- Procurement

  As mentioned at the beginning of this section, there should be a SLA or contract to abide by. For third-party offshoring, do we pay the offshore entity by instalments or a lump sum at the end of each project?

12.5.4. Outsourcing and offshoring executing

In this phase, the TPMP for transition or PMP for each subsequent development, together with SLA or contract will be implemented and managed. A set of automated tools should be deployed to oversee all aspects of the operation (i.e. transition, actual development), to address productivity risks and to conduct tangible benefit-cost analysis of offshoring. There should also be continuous improvement initiatives associated with the project on the transition process, development process and process change.

One item of paramount importance is quality. Quality metrics must be carefully captured, especially when the offshore entity is doing its development for the very first time.

12.5.5. Outsourcing and offshoring monitoring and controlling

It is important to monitor the progress of transition (knowledge and skills), progress of development, and performance of the outsourcing/offshoring entity. This is no different than running any project whereby the actual cost and schedule are compared with the baselines. Naturally, earned value analysis can be utilized to provide an objective and tangible assessment.

Measurement is difficult, as there is a tendency not to measure [7]. What criteria should be used by the originating entity to stop the transition or take back subsequent developments due to non-performance or unsatisfactory performance of the outsourcing/offshoring entity?

It is unfortunate, yet true, that a shot-gun wedding means an unusual business relationship. The outsourcing/offshoring entity can be a target of political attack in case anything goes wrong with the project. Do not forget that many peers are impacted locally by this outsourcing/offshoring venture and those who remain may feel bitter about this venture for a very long time.

12.5.6. Outsourcing and offshoring closing

This is the phase to put an orderly end to the transition and development, regardless of smooth transition, successful development, or abrupt termination. Lessons learned must be conducted. The expense of the transition and tangible saving in development after the transition are captured and reported. For details of benefit-cost analysis of offshoring, see Appendix F.

12.5.7. Outsourcing and offshoring lessons learned

There are too many advantages to mention for outsourcing/offshoring. Most notably, the tangible cost saving can be tremendous. For example, currently an engineer in North America can cost 3–5 times an engineer based in Asia, for equivalent competency. Companies can focus their energies on core functions while being relieved of ancillary duties [8].

There are disadvantages too: from now on, there is total dependence on the outsourcing/offshoring entity, somewhat like a heart-lung machine. There can be unsatisfactory consequences such as lost time, money, quality, and productivity [8]. There will inevitably be finger-pointing should anything go wrong. There are risks to confidential information. Within the local entity, attrition rate is going to rise as the smart rats are the first to leave a (perceived) sinking ship. Some may not stay to see the closing phase of the transition or development.

There is no right or wrong for the lessons learned. We leave discussions as open-ended questions for readers to develop courses of actions based on real-life situations. After all, is this act contradictory to our responsibilities to advance the profession while creating unemployment to society? Are there any non-tangible benefits and costs, or long-term impact to the profession and society as a whole?

12.6. Contract Law and the Project Manager

In the previous five sections, we use the word contract rather loosely. A contract is a legal instrument. It is technically a legal agreement between two people or two bodies sanctioned by law. There are certain steps that have to be followed. Failure to do so can land us in court and we can be charged with breach of contract.

12.6.1. Five key elements for a contract

For a contract to be legal, it must have five elements:

1. Offer made and accepted
2. Mutual intent to enter into the contract
3. Consideration
4. Capacity to contract
5. Lawful purpose

We now examine these step-by-step. Recall that the contract is established when the buyer has picked a particular seller with whom he wants to do business.

1. Offer made and accepted

The buyer makes an offer. The seller accepts that offer. This constitutes the first element of the contract. Contracts do not HAVE to be written, by the way, but it would be very unwise to enter into such a legally important position without having a written contract. The offer may be withdrawn by the buyer at any time if not accepted unless he clearly indicates it is irrevocable. It is also common to have counter-offers from the seller, where the roles are reversed. Acceptances on both sides must be clearly communicated. The contract may have an option which is irrevocable for a stated period of time. This is common. For example, in buying a house, the purchaser may put a percentage of the price of the house down to ensure that no one else can buy the house until a stated period of time expires. Options must be secured with something of value; hence the percentage down paid by the purchaser. In communicating between the buyer and the seller, both should specify the medium to be used. Email is the most likely medium but you could also use the postal system. The acceptance of the offer is effective when posted or when stamped by an email server. It is also important in this Internet age to realize that the governing law of the place where the buyer resides, will be used to define the legal position of both parties. That law takes precedence over any other laws of the land or the world, for that matter.

2. Mutual intent to enter into the contract

   Both sides must have clear intent to enter into the contract. Note that instruments such as letters of intent (LOIs) that do not specify details of the contract are not contracts. In the words of a former judge, the law does not recognize a contract to enter into a contract. Both sides of the contract must behave in good faith. If the seller offers to do the work, he cannot then renege on that acceptance. Similarly, the buyer cannot revoke her offer arbitrarily. Contracts can also be upturned if influence, duress, or fraud can be proven on either side. Because the contract without consideration is simply a gratuitous promise, it is not legally binding. As mentioned above, if the buyer promises to hold an offer open for a specified period, it must be accompanied by separate consideration.
3. Consideration

   Consideration means something of value in exchange for benefits of both parties. It could in the old days, be a red seal for example, sealed by a magistrate. Invariably, it is money that will be exchanged. That is one of the reasons you often hear of the person donating a building to a charitable organization demanding the payment of one dollar. That is because you need consideration to complete a contract. The contract will benefit both parties.
4. Capacity to contract

   You have to be in a legal position to contract. Minors, that is, people in Ontario under 19 years of age, cannot enter into a contract. Neither can drunks nor lunatics. Note that the word lunatic is a legal definition of a person who is not within his wits. A drunk can weasel out of the contract if one, the other party knows of the drunkenness and two, he repudiates the contract quickly.
5. Lawful purpose

   A contract is unenforceable if it is illegal or contrary to any statute. In gangster movies, they often make a joke about a gangster taking out a "contract" on somebody else; that is, they want them killed. That is not a legal contract. Other examples are violations of Workmen's Compensation; for example, paying somebody less than minimum wage, bid rigging, waiver of lien rights, etc. If the contract requires a licensed person, that person must be licensed for the contract to hold. For example, a person holding herself to be a licensed electrician who does work for a householder, and is not licensed, will not have her contract supported in the courts.

12.6.2. Fundamental definitions with contracts

Now we need some fundamental definitions that are used with contracts.

Enforceable. The contract is enforceable if the courts decide that all the elements of the contracts are in place. The opposite of enforceable is unenforceable.

Voiding. The term void is used to indicate that the courts have decided that a contract is unenforceable. A contract may be voided if a mistake was made, if there was misrepresentation, if there was duress, undue influence, unconscionability, frustration, or impossibility. Each one of these terms has a very narrow legal definition and a lawyer is needed to decide if they apply, not the PM. Duress, for example, means that actual violence has been directed against one of the contracting parties. Undue influence can lead to repudiation when the bargaining positions of the buyer and the seller are of unequal levels. An unconscionable contract is one that is so unfair, or oppressive or one-sided that the court would deem it offensive if it were to enforce it. Frustration means that something has happened that makes it impossible to complete the contract as intended. For example, during the First World War, any construction contract that depended on the availability of labourers in France or Germany would have been able to claim frustration because any man that was capable of doing manual labour would have been drafted into the Army to fight in that war. Impossibility happens when typically a force majeure (Act-of-God) event occurs. Examples such as earthquakes, forest fires, wars, hurricanes that make it impossible to complete the project may lead to a declaration of impossibility. The affected party is relieved from performing for a reasonable period of time.

Contract rescission. This is the act of setting aside a voidable contract. If one of the conditions listed above has happened, then we say the contract is in rescission. Contract rescission cannot be done unless the principle of *restitutio in integrum* is possible: that is, we can reset the financial situation so that nobody has lost any money.

Breach. Breach of contract is the failure of one of the two parties to perform her obligations according to the contract. Breach is a very serious situation. It means that somebody is going to pay a lot of money in penalties for not upholding her end of the contract. Once a contract is breached (e.g., non-payment), the other party does not have to perform from then on. There are several remedies for breach of contract:

- direct compensation
  The party that breached the contract compensates the other party on loss directly incurred due to its failure to perform.
- incidental/consequential compensation
  The party that breached the contract compensates the other party on loss indirectly or potentially incurred due to its failure to perform. This is up to legal interpretation and may or may not be granted.
- specific performance
  The party that breached the contract must perform what was being agreed in the contract.

Contracts may be amended. An amendment is a change to contract terms agreed on both parties. It is important to understand that a breach of contract is a loss for both parties. The only people that win when a breach occurs, are the lawyers. When something has gone wrong with the contract, an amendment is a much more intelligent way of coping with the situation, as opposed to putting the contract into breach. For example, if a buyer issues a "waiver", or agrees to waive certain terms and conditions, the seller does not have to provide the waived product or service.

Finally, we come to the concept of **estoppel**. Suppose we signed a contract to rent an apartment and we had agreed to pay our rent on the first of every month. The first of the month comes and we are a little short of money. So we go to our banker and explain the situation and she says "don't worry, as long as you pay by the end of the week, it will be okay." This happens several times. The next time it happens, the banker refuses and calls in the loan because we have not paid the rent on the first of the month. We can claim estoppel. This means that the banker cannot now enforce the terms of the loan because in the past, she did not rigorously do so. When running contracts, the conditions of the contract must always be enforced. If there is a condition that we should have enforced and we do not do so, we cannot revert to the original conditions at later date. This gives rise to the term forbearance, whereby the buyer gives right to non-compliance but does not waive the associated terms and conditions.

12.6.3. Contract interpretation and management

Common Law is arguably used the most in contracts. In North America, Common Law applies, except for the Province of Quebec in Canada and the State of Louisiana[1] in the US, where the Napoleonic Code (or similar variations) applies.

Common Law allows courts to create law and to use precedence (previous court cases of similar nature) to make judgement. The uniqueness of Common Law is that consideration must be given to both parties and a contract must benefit both sides. The Reasonable Man Standard, or what a person would reasonably do, is also used to make judgement.

The following rules are used to interpret a contract:

- parol evidence

Verbal agreement can constitute a contract. However, once there is a formal written contract in place, it cannot be amended by verbal means.

---

[1] The legal system in Louisiana is based on Civil Law with some Common Law influences. It is primarily based on the French and Spanish codes, and ultimately Roman code.

- rules of statements

  Numeric values (for example, monetary amounts) are always present in contracts. Should there be discrepancies between values written in words (e.g., one thousand dollars) and in numeric digits (e.g., $10,000), the written words take precedence.

- *contra proferentem*

  Contracts are interpreted against the drafter. The seller would typically draft the contract. Should there be contradiction or ambiguity within certain terms and conditions, the interpretation will be up to the buyer.

Project managers may consider the following five 'P's for contract management:

- people
- process
- performance
- price
- payment

PMs should also be aware of the rule of privity. Suppose a buyer enters a contract with the seller. The seller in turn subcontracts all or part of the job to a subcontractor. The rule of privity states (as shown in Figure 12.2) that there is legal contractual relationship between the buyer and seller, and a separate legal contractual relationship between the seller and subcontractor; but there is no contractual relationship between the buyer and subcontractor.

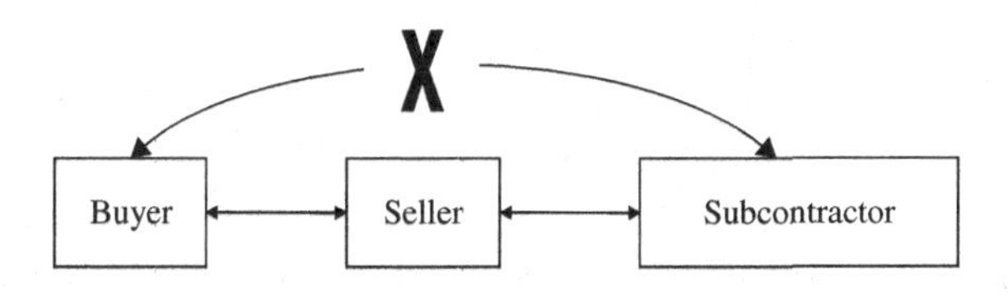

Figure 12.2. Rule of privity.

12.6.4. Mistakes in writing contracts and case study

What happens if one of the parties makes a major mistake in writing a contract? This is a vexing problem in contract law. If the mistake is a secretarial nature, the contract may be rectified if both parties agreed to the correction. As we mentioned previously, breaching a contract is of no advantage to either side. So if a simple change in the contract can solve the problem, that is the way to go. Over the last 40 years, a significant number of cases have aided greatly in the clarification of contract law. Generally speaking, the courts will not overturn a bad business deal. However, if the contractor makes an honest mistake, which does not go to the heart of the contract, it is highly likely the courts will allow a modification of the contract to rectify the mistake.

In Bell River versus Kaufman, Ontario Report, 1977 ([9], see also [11], p. 117), a case is described where a contractor submitted a bid on a contract. He realized after he submitted the bid that his secretary forgot to transfer some figures from an adding machine to a summary page. He attempted to withdraw from the contract. The plaintiff refused and held it for 30 days and then attempted to accept the contract. He accepted another higher bid and sued for the difference. The judge in that matter ruled for the contractor, noting that there was a clerical error, not an error of professional judgement. He might have added the error did not go to the heart of the contract. In making the judgement, the judge wrote that "a just and reasonable man will not assist in profiting from another's mistake. A substantial mistake must occur for the breach to be permitted. The proof for that mistake must be strong and convincing." Bell River was the watershed concerning mistakes for many years.

In 1979, a landmark case in contract law occurred. The case was called Ron Engineering (Ron) versus the Queen ([10], see also [11], p. 118). A mistake superficially similar to Bell River occurred. The contractor submitted a tender along with a bid deposit of $150,000 to the Water Resources Commission of North Bay. The contractor had omitted an amount of about $750,000 out of the total bid of $2,748,000. She later realized her mistake and attempted to withdraw the submission before the opening of tenders. But the Water Resources Commission refused, which was their right. Ron informed the plaintiff again an hour after the opening verbally and by telegraph that they wanted to withdraw the tender. The Water Resources Commission refused to annul the bid and indeed awarded the contract to Ron, they being the lowest bid by far. There was no doubt about the error. The next closest bid was over $3.3 million. Ron sued to get its bid deposit back, even though the RFP clearly stated that the Water Resources Commission did not have to return the deposit if the contractor did not go to contract. The original presiding judge ruled in favour of the Water Resources Commission. Ron appealed to the Superior Court of Ontario, quoting Bell River and won on the appeal. It was further appealed to the Supreme Court of Canada (SCC) and overturned again, the SCC ruling against Ron. In stating its decisions for the overturning, the SCC established the concept of what we now call Contract A and Contract B. In the writing of that judgement, the judges stated that an advertisement for tenders is an offer that is accepted when a tender is submitted and a contract is formed that precludes the contractor from withdrawing her bid. This is the doctrine of Contract A and Contract B. Contract A puts the onus on both the buyer and the seller to obey by the conditions listed in the RFP. The buyer sets the rules in the original RFP. The seller responses to the RFP, according to the rules as best she can. Those rules cannot be changed. If they are changed for any reason, that is grounds to breach the contract. Once a contractor is selected, then we enter into Contract B which is the normal contracting process that we have seen before. As long as there is no major change to the contract terms which go to the heart of the contract, then the contract can be formed in the normal fashion. The bottom line for the PM is that she must understand what the terms of the RFP are. Those terms are legally binding on both the buyer and the seller. There are several cases that are referred to in Marston's book on Law for Professional Engineers [11], if you wish to examine this in more detail.

12.6.5. Contract termination and case study

The legal term for contract termination is discharging. Contracts can be discharged for several reasons: performance, mutual agreement to discharge, discharge pursuant to express terms, discharge by frustration, or discharge by breach. In the first case, the contract is discharged because of the positive outcome of a good contract. The contract was performed well to the buyer's satisfaction. Both parties agree to discharge the contract and get on with their work. In the second case, the agreement to discharge mutually means that both sides have decided it is time to terminate the contract and typically that will be done with the execution of the doctrine of *Quantum Meruit*, which is Latin for "as much as is reasonably deserved". In legal terms, it refers to paying for expenses out-of-pocket that may have occurred in the mutual discharging of the contract. For example, the seller may have purchased some materials in anticipation of continuing the work. When the contract is mutually discharged, he would naturally want to be reimbursed for those expenses. The third type of discharging is to pursuant to express terms, for example, one of the parties goes bankrupt. That would be a reason to terminate the contract and the conditions of the termination should be listed in the contract itself. The fourth type is discharge by frustration, which we have already discussed. That occurs when some natural event occurs that makes it impossible to continue the execution of the contract, such as scarcity of people to do the work. Finally discharges by breach occur when somebody has fundamentally broken a condition which goes to the heart of the contract. In this case, we are going to court with our respective lawyers and the judge will decide how the contract should be wrapped up.

It is the case in the drafting of the contract that the drafter of the contract will try to protect herself wherever she can. Often exemption clauses are included which purport to shield one side or the other from undue expenses. You may find a term such as "limiting the contractor's liability to $10,000". Those exclusions do not prevail if it can be shown that the contractor did something that was not professional. It is important for the PM to realize this when she executes her work. You must be prepared to answer the lawyers' question "did you do the work in the way that a reasonable practitioner of this field would have done?" If you have exercised due diligence in the execution of your work, you will prevail in court; but if you did not, you will not.

Here is a spectacular example of that; the case of Harbutt's Plasticine versus Wayne's Tank and Pump, an English case from the House of Lords, 1970 ([12], see also [11], p. 155). The contract for the design and installation of storage tanks for stearine was at issue. Part of the contract involved designing a plastic pipe line wrapped with electrical heating tape. It was to liquefy the stearine so it could flow from one point of the plant to another. The contract also contained a clause limiting the contractor's liability to $4000. The pipe sagged, cracked, and the stearine leaked out, catching fire and burning the place down. The actual loss was $300,000. The contractor was ruled to be in fundamental breach of the contract. The reason that he was in fundamental breach is that he did something really unprofessional in the construction of the pipe. Therefore, this exclusion clause would not shield him from responsibility. The PM must keep this in mind at all times. Keep adequate notes showing that you have followed proper project management procedures. If you are following the PMBOK, for example, then this is an excellent way to document that you are exercising due diligence in the running of your project.

12.7. Multiple Choice Questions

1 When a project manager places a purchase order for a piece of equipment, it represents:
   (a) Commitment
   (b) Expense
   (c) Cash out-flow
   (d) Capital investment

2 Which of the following are frequently used tools to Plan Procurements:

(a) Make or buy analysis, expert judgement, contract type selection
(b) Contract type selection, bidder conferences, expert judgement
(c) Expert judgement, audits, bidder conferences
(d) Make or buy analysis, contract type selection, weighting system

3 During the contract close-out, the project manager needs to document the:

(a) Formal acceptance
(b) Statement of work
(c) Payment schedule
(d) Change control procedure

4 Which one of the following is not a process for Project Procurement Management:

(a) Plan procurements
(b) Conduct procurements
(c) Request seller responses
(d) Administer procurements

5 Which one of the following is not a competitive contract method?

(a) IFB
(b) RFP
(c) Reverse auction
(d) RFI

6 Which one of the following contract types is of higher risk to the seller?

(a) FPIF
(b) T&M
(c) CPIF
(d) CPFF

7 You have been contracted to implement a project. The estimated cost of the project is $1,000,000. The contract type is CPIF with the agreement that any cost savings will be split between you (20%) and the buyer (80%). If at the end of the project, the cost for the project is $800,000, how much will be awarded to you?

(a) $800,000
(b) $840,000
(c) $1,000,000
(d) $960,000

8 What is a SOW?

(a) Narrative description of a product or service to be provided by a seller to the project
(b) Narrative description of a product or service to be provided by a buyer to the project
(c) Narrative description of a product or service that sanctions the project
(d) Project plan of a product or service to be provided by a buyer to the project

9 You are looking to buy a project management tool with MTTF of higher than 5 days. What standard of evaluation have you imposed?

(a) Absolute
(b) Minimum
(c) Relative
(d) Weights

10 Which one of the following dispute resolution outcomes is final?

(a) Negotiation
(b) Mediation
(c) Arbitration
(d) Litigation

References

1. National Academy of Engineering. *The Offshoring of Engineering*. The National Academies Press, 2008.
2. Contractor, F. J., Kumar, V., Kundu, S. K. and Pedersen, T. *Global Outsourcing and Offshoring*. Cambridge, MA: Cambridge University Press, 2011.
3. McIvor, R. *Global Services Outsourcing*. Cambridge, MA: Cambridge University Press, 2010.
4. Healey, M. Fast, Cheap, Good: Pick Two. *Informationweek*, April 2011.
5. Putnam, L. H. and Myers, W. *Measures for Excellence*. Prentice Hall, 1992.
6. Wu, T., Blackhurst, J. and Chidambaram, V. A model for inbound supply risk analysis. *Computers in Industry*, 57, 2006.
7. Preston, R. Outsourcing Customers Send Some Mixed Signals. *Informationweek*, February 2007.
8. Haugen, D. M., Musser, S. and Lovelace, K. *Outsourcing Opposing Viewpoints*. Greenhaven Press, 2009.
9. Bell River Community Arena Inc v. W. J. C. Kaufman Co. *et al.* (1977) 15 O.R. (2d) 738.
10. Ron Engineering *et al.* v. The Queen in right of Ontario *et al.* (1979), 24 O.R. (2d) 332.
11. Marston, D. L. *Law for Professional Engineers*, 4th Edn. McGraw-Hill, 2008.
12. Harbutt's Plasticine Ltd. v. Wayne's Tank & Pump Co. Ltd., (1970) IA11 E.R. 225.

CHAPTER THIRTEEN

Stakeholder Management

Project stakeholder management refers to the processes that identify stakeholders, analyze their expectations and develop strategies for their engagement. It focuses on continuous communication with stakeholders and their satisfaction.

13.1. Stakeholder Management Initiating

In this phase, the most important task is to identify the stakeholders. Who are the stakeholders? They are individuals or organizations whose interests are affected by the project. They may exert influence over project objectives and outcomes. There are positive or negative stakeholders. Some examples of stakeholders include the project manager (PM), customers, users, performing organizations, project team, project management team/office, sponsor, etc. Their authority (power), interests (concerns), expectations, involvement (influence), and impact should be documented. Strategy should be established to maximize positive influence and mitigate potential negative impacts. Figure 13.1 illustrates the concept [1].

The major output of this phase is the Stakeholder Register. This document contains the name and interests of all the identified stakeholders. The register should contain:

- identification: information concerning the particular stakeholder, such as name, title, email address, office address and telephone number.
- assessment information: what is the interest of the stakeholder in this project? What is her power in the organization? What requirements will the stakeholder be able to qualify?

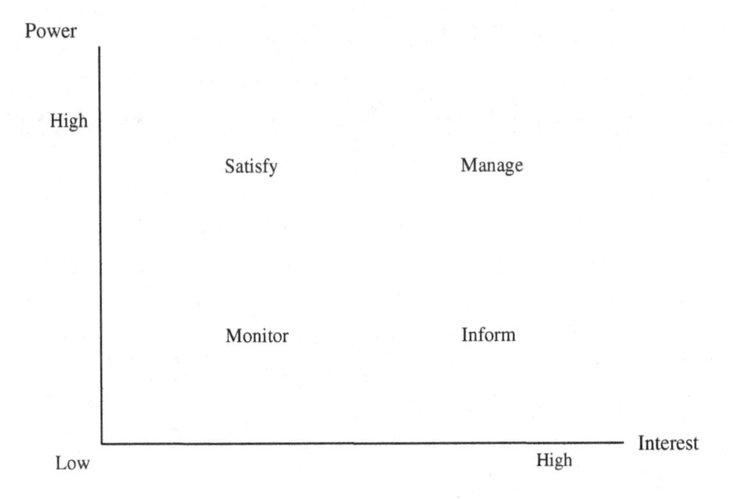

Figure 13.1. Stakeholder power and interest grid.

- stakeholder classification: is the stakeholder internal/external, supporter/neutral/resistor, or is there any other meaningful classification that you can make?
- reporting frequency: how often must we communicate with the stakeholder, and what must we communicate?

The Stakeholder Register is the final output of this phase.

13.2. Stakeholder Management Planning

This process details how current and desired levels of stakeholder engagement are analyzed. The levels of engagement are:

- unaware
- resistant
- neutral
- supportive
- leading

The levels of engagement may change according to project situation. This process also identifies the interrelations among stakeholders.

13.3. Stakeholder Management Executing

This process manages stakeholder engagement by communicating and working with the stakeholders. Concerns should be addressed before they mature into issues. Future problems are anticipated to assess project risks. The goals are to increase support and minimize resistance from stakeholders.

13.4. Stakeholder Management Monitoring and Controlling

This process ensures the comprehensive identification and listing of new stakeholders, reassessment of current stakeholders and removal of stakeholders no longer involved in the project. It also monitors changes in stakeholder interrelations and controls stakeholder engagement in order to adjust the stakeholder strategies if needed.

References

1. Project Management Institute. *A Guide to Project Management Body of Knowledge* (PMBOK), 5th Edn. PMI, 2013, p. 397.

CHAPTER FOURTEEN

Ethics and Engineering Professionalism

14.1. Topics for this Chapter

We have already introduced the topics of Ethics and Professionalism in Chapter 2. We now discuss these in detail. Ethics is the study of morals by which people draw up sets of imperatives or principles of right conduct. The morals of an individual are the ways in which the imperatives are applied in action. The professional engineer decides what is the most appropriate action to follow for the client by virtue of one's training and experience. In this chapter, we will cover the Code of Ethics and Professional Conduct of a Project Management Professional (PMP), Code of Ethics of a Professional Engineer, prevention of sexual harassment in the workplace, management triples, and basic human fundamentals. We are also going to present several case studies as exercises at the end of the chapter. While we are fully aware of plagiarism, we believe that presenting the exact wordings for certain code of conduct and ethics, as well as legal terms would be appropriate for this chapter. We will make reference to these in their first occurrence only.

14.2. Code of Ethics and Professional Conduct of a Project Management Professional

A Project Management Professional (PMP) is committed to doing what is right and honourable at work, at home, and in service to the global project management community. One has to support and adhere to the responsibilities described in the Code of Ethics and Professional Conduct [1] in both the professional and volunteer roles. That code covers three main areas:

- responsibilities to the Profession,
- responsibilities to the Public and Customers, and
- administration of the Code of Ethics and Professional Conduct.

The code also affirms the four values of its foundation: responsibility, respect, fairness, and honesty. Responsibility is the duty to take ownership of our decisions and actions. We do what we say we will do. When we make mistakes or omissions, we take ownership and make corrections, and we accept accountability for any consequences. Respect is our duty to show a high regard for ourselves, others, and the resources entrusted to us. We do not engage in behaviours that others might consider disrespectful. We conduct ourselves in a professional manner and directly approach those with whom we have a conflict or disagreement. Fairness is our duty to make decisions transparently and act equitably — without prejudice, favouritism, and discrimination. Honesty is our duty to understand the truth and act in a truthful manner. We make commitments and promises, implied or explicit, in good faith.

14.2.1. Responsibilities to the profession

The responsibilities include compliance with applicable laws, regulations, and standards in the state/province and/or country, and compliance with all organizational rules and policies. One accepts assignments that are consistent with one's background, experience, skills, and qualifications. One must provide accurate and truthful representations concerning all information related to PMI certification and also maintain and respect the confidentiality of the contents of the PMP examination. One must also hold others accountable and report violations of the Code of Ethics and Professional Conduct. Finally, one is responsible for advancement of the profession.

14.2.2. Responsibilities to the public and customers

The responsibilities include accurate and truthful representations to the public in advertisement and public statements, as well as satisfactory execution of the scope and objectives of professional services, based on the best interests of society, public safety, and environment. In addition, confidentiality of sensitive information must be maintained and respected. A conflict of interest must not exist that could compromise our integrity or professional judgements, as well as the legitimate interests of clients and customers. One shall not exercise the power of expertise or position and shall not engage in dishonest behaviour, in offering or accepting inappropriate payments, gifts, or other forms of compensations for personal gain.

14.2.3. Administration of the code of ethics and professional conduct

While one is to act in an honest and ethical manner, any conflict of interest must be disclosed to related parties or the stakeholders. One then refrains from decision-making or influencing the outcomes. Each value of the code's foundation includes aspirational standards and mandatory standards. Aspirational standards are expectations that one strives to uphold as a professional. Mandatory standards establish firm requirements that limit or prohibit certain behaviour. Concerning ethics violation, one must co-operate with the PMI in the collection and disclosure of information. Violation of mandatory standards is subject to disciplinary procedures before PMI's Ethics Review Committee.

14.3. Code of Ethics of a Professional Engineer

We discuss the Code of Ethics of a Professional Engineer [2] based on the Professional Engineers of Ontario (PEO). The Code of Ethics is a guide to the professional as to one's conduct in relation with the clients, employers, colleagues, business contacts, society, etc. Other engineering bodies have similar but not identical Code of Ethics. All of them include at least the following:

- fairness and loyalty to associates, employers, clients, subordinates, and employees;
- fidelity to public needs; and
- devotion to high ideals of personal honour and professional integrity.

A professional engineer accepts the Code of Ethics as a series of guidelines designed to assist herself as a practitioner in conducting a professional engineering practice. She shall undertake only such work as she is competent to perform. She shall conduct herself toward others with courtesy and good faith. She is obliged to give proper credit for engineering work. She shall discourage untrue, unfair, or exaggerated statements. She shall not accept any engagement to review the work of others except with their knowledge.

A professional engineer ensures full disclosure to the affected parties is made prior to his engagement. He shall make effective provision for the safety and health of the public, for meeting lawful standards, rules, and regulations. He has a clearly defined duty to society and to regard public welfare as paramount. He shall maintain confidentiality of sensitive information such as business affairs, trade secrets, technical methods of a proprietary nature, and processes of an employer. He shall enhance the public regard for the engineering profession and extend the effectiveness of the profession.

Engineering notes may be used in the future as a defense in the case of legal action. An engineer should always keep notes of her professional activities in ink and use a bound stitched notebook to record them.

14.4. Preventing Sexual Harassment in the Workplace

This is perhaps the most important mandatory training in most academic institutions and industrial organizations. Harassment is abusive, harassing, or offensive conduct in a verbal, physical, or visual manner. Sexual harassment [3,4] refers to inappropriate comments, actions or conduct of a sexual nature to be known as unwelcome. Here is an incomplete list of examples:

- gender-related comments about an individual's physical characteristics or specific gender
- physical contact such as touching, hugging, patting, rubbing, or pinching
- restraining or blocking the path of a co-worker
- leering or inappropriate staring
- demands for dates or sexual favours
- offensive jokes, humor, or comments of sexual nature
- display of sexually offensive pictures, posters, calendars, or other materials
- questions or discussions about sexual activities or orientation

Organizations must not tolerate sexual harassment. You have the right to let others know that their behaviour is offensive and demand them to stop, if the harassment is directed toward you. If you see others being harassed, you must report what you saw. Organizations want you to report any sexual harassment complaint or activity to your supervisor or manager, or any member of management, or human resource representatives. Organizations will ensure that the complaint is treated seriously, with sensitivity and with confidentiality, and that no retaliatory actions are taken against you as a result of filing a good faith complaint.

Sexual harassment is not always recognizable because our individual perceptions differ. Each person brings one's viewpoint into focus as interpretation of the behaviour of the others. Sexual harassment does not have to be intentional.

Although sexual harassment affects women primarily (i.e., harassed by men), men can be harassed too! In addition, men can harass men, women can harass women. Note though that many studies have shown that 19 out of 20 harassers are male! Offenders can be within the same organization or they can be customers, vendors, suppliers, etc.

As the leader of a team, how do you treat a sexual harassment complaint?

1) First, obtain the names of the (potential) offenders; then understand the specific actions/behaviour that the complainant found offensive.
2) Obtain the names who may have observed (i.e., witnessed) these actions/behaviour.
3) Obtain the names of people who may have had similar encounters in the past.
4) Keep the situation confidential, document and repeat what you understood to the complainant, and advise the victim of your next steps.
5) Exercise due diligence in investigation, engage other persons in the organization designated to assist in such circumstances, take appropriate actions (including dismissal of the offenders) to correct the situation.

14.5. The Management Triples

We assume that a regular engineering student who graduates from any accredited engineering school is going to possess adequate technical skills and knowledge for his/her engineering domain (e.g., chemical, civil, computer, electrical, mechanical, nuclear, software) [5]. In today's world (at school, at work or at home), it is a "given" that one must multi-task. Can one "manage oneself"? After entering the industry for 2–5 years, one expects to be promoted to a senior engineer leading a couple of junior engineers

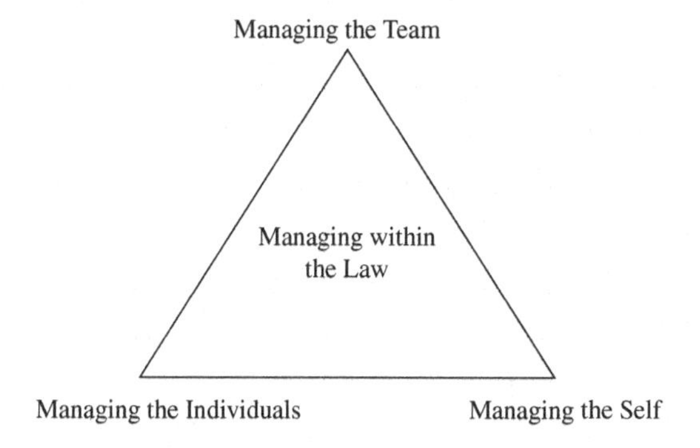

Figure 14.1. The management triples.

and/or student engineers. Can one "manage the individuals"? After 5–10 years in the industry, one is expected to undertake a team leader or management role. Can one "manage the team"? Most importantly, one must always "manage within the law". We illustrate the management triples in a triangle as shown in Figure 14.1, with the key (the fourth one) written in the middle.

Using the management triples as foundation, one can recognize and take ownership for the key role that a manager plays in ethics — as a positive activist and role model for the organization's value and for taking appropriate measures in the event of violations.

The code of conduct is applicable to many, but not limited to:

- Co-workers

  Demonstrate constant respect to all co-workers, maintain a safe and healthy work environment, and exercise zero-tolerance for harassment and discrimination.

- Consumers/Customers

  Comply with applicable laws and regulations in creating safe and quality products. Tell the truth about products and services. All marketing and advertising shall be accurate and truthful. Protect customer information that is sensitive, private, or confidential.

- Business partners
  Treat partners and suppliers fairly. Do not engage third parties (i.e., agents) to perform illegal acts. Work to make the standards of joint ventures compatible with your own.

- Shareholders
  All financial books, records, and accounts must accurately reflect transactions and events, and conform to required accounting principles. Safeguard all business information, assets, and proprietary information. Do not trade on inside information.

- Competitors
  Compete aggressively but with integrity at all time. Gather competitive information legally.

- Communities
  Be a good corporate citizen. Protect the environment by complying with applicable environmental laws in all countries.

- Governments
  Comply with the law. Bribes or kickbacks are never acceptable.

For ethical analysis, one has to first fully understand the facts, then refer to the Code of Conduct and the key beliefs. Finally, consider what is right for all involved.

14.6. Basic Human Fundamentals

There is no silver bullet to successful management. Taking relevant training and following PMBOK or other standards will surely lead to running projects with less risks and higher probabilities of success. Nevertheless, there is no replacement to some basic human fundamentals. Confucius taught about the Four Supports and Eight Virtues as human fundamentals. The Four Supports are Propriety, Justice, Integrity, and Self-Respect. These, along with Filial Love, Brotherly Love, Fidelity, and Trustworthiness, form the Eight Virtues. Of course, it is not entirely necessary to attain "sainthood" for successful management. In today's "hi-tech" world, we

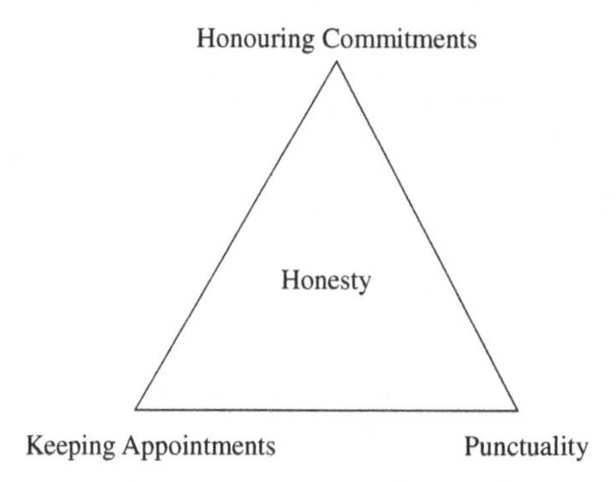

Figure 14.2. Basic human fundamental triples.

emphasize punctuality, keeping appointments, honouring commitments, and most importantly honesty; refer to Figure 14.2 (which is analogous to Figure 14.1). We believe that these form the foundation for the management triples, which in turn are the backbones for ethics and professionalism.

We conducted a survey (based on a real-life example) to study the impact of technology on the human fundamental triples for the upcoming generation. We like to tease our students by saying that two individuals, while walking inside a shopping mall, were so busy texting each other to find out where they were that eventually they bumped into each other "heads on". While technology has enabled communications easily, would over-reliance on technology create "the tool and a fool" problem? Is the upcoming generation seeing last-minute changes via technology to appointments and commitments as part of the norm?

Our survey was targeted toward the upcoming generation, specifically the undergraduate and graduate students, who would attain their professional designations shortly after graduation. Although participation is entirely optional, we received 118 enthusiastic responses, with just a few opt-outs. There were 10 multiple choice questions that would take less than 5 minutes to answer. There were no right or wrong answers. Instant responses were the best answers. Appendix E listed the questions. The answers ranged from ratings of very likely (5), somewhat likely (4), neutral (3),

Table 14.1. Results of survey.

Questions	Scores (%)	Questions	Scores (%)
1	51	6	65
2	85	7	76
3	68	8	87
4	68	9	58
5	60	10	64

somewhat unlikely (2), to very unlikely (1). The results were normalized and presented as percentages ranging from very likely (100%) to very unlikely (0%).

We are not going to repeat the questions, but merely highlighting our interpretation of the results in Table 14.1.

Q1. In hectic and adverse moments, half of the group will forget about their commitments. The result is rather astonishing! Note that these are just answers on paper. In real-life situations, more will forget! Note how we presented the importance of this commitment: deemed unimportant to the provider but how about the requester?

Q2 and Q3. Luckily, a very high percentage will notify the other party in advance and a high percentage will honour the make-up session.

Q4 and Q6. These questions bring up the "tool and a fool" scenario. There was one written comment asking whether the other party actually received the email. One does not know, and what does one do about it?

Q5 and Q10. These questions follow the Porter assessment, when one is being challenged (or cornered). It appears that the upcoming generation does not have a high level of tolerance.

Q7 and Q8. We feel relieved by the high percentages for these answers — feeling guilty and making up the omission.

It appears that the majority of the group is still observing the basic human fundamentals of courtesy. However, we realize that we have only solicited answers on paper, in an artificial environment. In real-life situations, the actual circumstances can influence the scores. The bottom-line is: how do you manage these professionals? What are their approaches to management and how do they manage in the future?

14.7. Chapter Summary

We have discussed several important topics related to ethics and professionalism. While a lot of these go back to common sense, it takes time and practical experience for one to mature through one's career.

14.8. Case Studies

Case 1

Dawn was a graduate from the University of Eastern Ontario (UEO) and is now working for a large telecommunications company. She has an excellent relationship with UEO and constantly keeps in touch with several professors. After graduation, she was invited to provide guest lectures in many occasions. When UEO started the Software Engineering program, Dawn was invited to co-teach Software Engineering courses with the professors regularly. She was later appointed as an Adjunct Professor at UEO. The company approved her collaboration with UEO. In addition, the company sponsored her time and expense to teach at UEO one day every two weeks, and to conduct research with the professors and graduate students. As part of Dawn's employment agreement with the company, the company has the right to any intellectual property and patent during her term of employment.

Q1. What is the legal status of her inventions with UEO's professors and students?

Jill is a good friend of Dawn and runs a startup telecommunications company. Jill's company was formed based on her patented idea related to security in wireless communications. Jill is actively looking for funding to develop products based on this idea. Dawn introduced Jill to UEO. A UEO professor has expertise in this area and so does Dawn. They applied for a government research grant to further the work. Jill was so thankful and invited Dawn and the professor to her company's Board of Advisers. Later on, the research grant was approved and students were developing prototypes under the joint supervision of Dawn and the professor. Jill was so happy that the products were almost usable by her company. She decided to grant some shares of her company to both Dawn and the professor. Dawn was a bit puzzled — she is now developing prototypes that will turn into products for Jill's company. The products may actually be in direct competition with some products of her employer. To make things more complicated, she is now an adviser for Jill's company (a potential competitor for her employer) and she is going to own shares of this competitor too.

Q2. Should Dawn accept the invitation to the Board of Advisers? Should Dawn accept the shares of the startup company?

After a year, with some products to showcase, Jill became very ambitious and started to solicit additional funding or investment opportunities from large telecommunications companies, including Dawn's employer. Dawn's boss came to her one day and asked her to join the committee and judge whether this startup company's patent and products were technically sound, and whether to put a substantial investment into this startup company.

Q3. What should Dawn do?

Case 2

You are the Manager of Facilities and found out that hazard materials are by-products due to cutback of the manufacturing process. The disposal is very expensive. An unknown contractor offers 30% lower price than the current contractor for disposal. The contractor believes that its method for disposal meets local legal requirements.

Q1. What would you do?

Q2. If a joint venture partner also uses this contractor, what would you do instead?

Case 3

A product is supposed to be shipped shortly and certain quality standards have not been met. Chances are that most customers will not encounter these flaws under normal usage. Some co-workers suggest that the test results be altered so that we still ship the product on time.

Q1. What would you do?

Q2. What if your manager tells you to omit these test results?

Q3. What if the customer agrees to take the product regardless?

Case 4

Your company was located in a rather old commercial building. The air conditioning pipes ran above the ceiling of the second floor. In a hot day, condensation along the pipes would cause water dripping down certain areas of the ceiling. The janitor chose to collect the dripping water by placing plastic buckets above the removable ceiling tops.

You were the manager of a test team of five. The whole team came to work overtime on a Saturday in order to fight schedule slippage. Sihe was part of the test team and she came to work too. Her husband dropped her off and left the building. Later on, she went to the ladies restroom. While she was there, a bucketful of water splashed down from the ceiling and wet her head and clothing completely! (The janitor forgot to empty this bucket regularly and the bucket became full and heavy, and eventually fell through the ceiling.) Sihe rushed out of the restroom, yelling and crying, but did not appear to be hurt. The team surrounded and comforted her, and another team member, Alan, reported this to you.

Q1. What would you do (to her, to the team, to building management, to your own senior management/executives, etc.)?

Alan offered to drive Sihe home. After cleaning herself up, Sihe came back to work.

Q2. What would you do then?

The team relaxed a bit after making good progress with the test project. They started teasing Sihe on the incidence. Sihe was smiling and laughing initially. (She is Islamic.) Alan suddenly said, "Ha! She got baptized by Holy Water from Heaven." Everybody laughed and clapped their hands. Sihe was not pleased.

Q3. What would you do?

References

1. Project Management Institute. *Project Management Institute Code of Ethics and Professional Conduct.* PA, USA: PMI, Inc., 2006.
2. Professional Engineers Ontario. *Professional Engineers Ontario Code of Ethics.* Canada: Professional Engineers Ontario, Ontario, 1990.
3. Ontario Bill 168, Occupational Health and Safety Amendment Act (Violence and Harassment in the Workplace). Ontario, Canada, 2009.
4. Policy on Preventing Sexual and Gender-Based Harassment. Ontario Human Rights Commission, Ontario, Canada, March 2011.
5. Bennett, J. M. and Ho, D. Project Management as an Essential Core Competency in Engineering. Annual Canadian Engineering Education Association (CEEA) Conference, St. John's, Canada, June 2011.

APPENDIX A

Waterfall Model for Software Development

The Waterfall Model for software development is represented by Figure A [1].

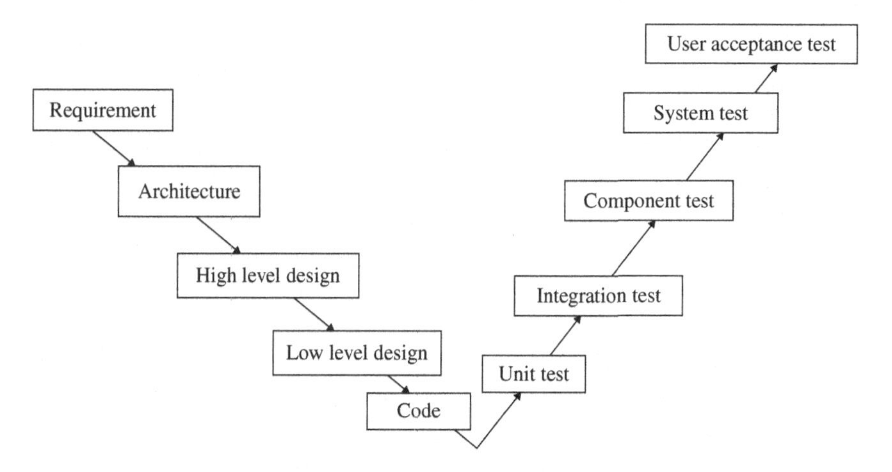

Figure A. Waterfall model for software development.

The waterfall model is the fundamental model based on which other software development models (e.g., prototype, spiral, iterative, object-oriented, agile/scrum, etc.) have been created. We are not discussing the phases within the model individually as the names are intuitive. Each phase has an entry, validation, and exit criteria, and a set of tasks to complete. Imagine there is water flowing down the rectangular boxes (phases). A box at the higher level must be filled with water first before the box below can be filled. By the same analogy, each phase must be entered by satisfying the entry criteria, executed by finishing a set of tasks, tested by passing validation, and exited by satisfying the exit criteria. Then, the next phase can be started.

Note that the unit, integration, component, system, and user acceptance test phases are indeed boxes located further down the path. We tilt the line upward in the diagram to give an overall V-shape in order to illustrate the mapping of various tests to their respective development phases. For example, unit test validates the code and low-level design; integration test validates the high-level design; component test validates the architecture; system test validates the requirements; and user acceptance test acts as the final touch for the end user to accept the product.

References

1. Available at: http://www.waterfall-model.com/v-model-waterfall-model/.

APPENDIX B

IEEE PMP Sample

General

Title: Spent Nuclear Radiation Fuel (SPENTFUEL)

Signature: (electronic)

Change history

Version	Date	Author	Notes
V1.0	yyyy/mm/dd	JMB	Initial Draft
V1.1	yyyy/mm/dd	DHo	Revision after formal review

Preface

This is the Project Management Plan for SPENTFUEL, which is a nuclear engineering software project to develop a database to record the location, agency in charge, burial details, and amounts, of all of the spent radiation burials in North America (NA). This database would be accessible to all nuclear plants in NA and of course, to appropriate regulatory officials. The idea is that if we ever find a permanent solution, all of the spent radiation could be collected and disposed of. SPENTFUEL will tell us where they are located.

TOC

(self-generated)

List of Figures

(self-generated)

List of Tables

(self-generated)

B.1. Overview

B.1.1. Project summary

B.1.1.1. *Purpose, scope and objectives*

The purpose of SPENTFUEL is to provide a central repository containing the location and amount stored for all nuclear burial sites, thus providing easy access to the spent fuel. The scope of SPENTFUEL then is to build this database for coverage in NA.

B.1.1.2. *Assumptions and constraints*

Funding will be available at the appointed milestones. Appropriate software skill-sets are available for the team selection. Members of the functional units are available when needed. Software and hardware platforms will be available when needed. If training is necessary, it will be provided and the schedule will be adjusted accordingly.

B.1.1.3. *Project deliverables*

This software project will deliver two major modules: control module and database module. The software as a whole will be delivered with documentation and evidence of test completion. Stakeholders and end users will be shipped a prototype after 12 months and the final product after 18 months.

B.1.1.4. *Schedule and budget summary*

This project is to be completed in 18 months for the final product, with an interim checkpoint in 12 months for a prototype. The effort internal to Vacux will be less than 150 staff-months. The monetary expense will be less than $100,000 for software, hardware, tools, travel, etc.; and $200,000 for the outsource/offshore contract.

B.1.2. Evolution of the plan

The PMP will evolve during of the project. Updates are mandated at each major milestone, or whenever significant scope change is incurred.

B.1.3. Charter

Already approved, see archive in project library.

B.2. References

Vacux corporate software development process (see company intranet web page)

Vacux corporate charter, PMP, schedule, quality templates (see company intranet web page)

B.3. Definitions

CNA: Canadian Nuclear Association

CPI: Cost Performance Index

DND: Department of National Defense

NA: North America

PMP: Project Management Plan

SPENTFUEL: spent nuclear radiation fuel

SPI: Schedule Performance Index

B.4. Project Organization

B.4.1. External interfaces

Canadian Nuclear Association (CNA)

Department of National Defense (DND)

Outsource/offshore entity for database development

Other stakeholders

End users

B.4.2. Internal structure

Management team: consisting of the project manager and two team leaders

Database professionals: consisting of database experts

Data communication professionals: consisting of TCP/IP communication experts

System test team: consisting of test experts

Quality team: consisting of quality consultants

B.4.3. Roles and responsibilities

Management team: accountable for the entire project, responsible for communication with CNA, DND, outsource/offshore entity, stakeholders, end users; and overseeing development of the entire project

Database professionals: overseeing outsource/offshore database module development

Data communication professionals: responsible for in-house control module development

System test team: setting up test lab and conducting system test of SPENTFUEL

Quality team: responsible for deriving quality metrics for the project and auditing the quality of the development process

B.5. Managerial Process Plans

B.5.1. Start-up plan

B.5.1.1. *Estimation plan*

The analogy method is used for deriving the high-level estimation for the time being. Bottom-up estimate will be provided by individual staff when various teams are formed.

B.5.1.2. *Staffing plan*

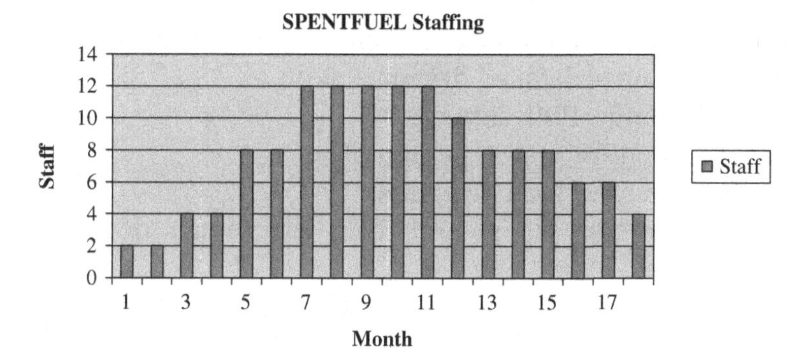

B.5.1.3. *Resource allocation plan*

Staff will be acquired and allocated to the management team, database team, data communication team, system test team, and quality team. Initially, one staff (working part time) will be nominated from each team to form a core group that works out the plan for the project. More staff will be assigned to the project in later phase per schedules of individual teams.

B.5.1.4. *Project staff training plan*

Existing project staff will be trained in accordance with one's personal development plan (subject and date per individual's plan).

New staff will be given training for the required expertise (e.g., management, database, TCP/IP).

Diversity training will be provided for the entire team at the beginning of the project.

Ethics training will be mandated annually per Vacux corporate policy prior to year end.

B.5.2. Work plan

B.5.2.1. *Work activities*

- Start + 3 months: requirements sign-off
- Start + 6 months: detailed design sign-off
- Start + 7 months: PMP sign-off
- Start + 12 months: prototype sign-off
- Start + 15 months: local acceptance test
- Start + 18 months: complete delivery of product, documentation, etc.

B.5.2.2. *Schedule allocation*

There will be separate schedules for individual teams. These schedules will be merged into the master schedule to be managed by the project manager and 2 team leaders.

B.5.2.3. *Resource allocation*

Management team: 3

Database team: 2

Data communications team: 12

System test team: 5

Quality team: 2

B.5.2.4. *Budget allocation*

Software (Windows 7 32-bit and 64-bit versions; Linux 3.5; Snow Leopard 9.1; Java version 7.1; other development tools): $20,000

Hardware (including lab setup): $50,000

Travel: $30,000

Outsource/offshore contract: $200,000

B.5.3. Control plan

B.5.3.1. *Requirements control plan*

Requirement and scope changes can be submitted to the Change Control Board of the project for consideration. The board meets biweekly and will table all change requests for consideration. Turnaround time for decision making will be three weeks from the date of submission. Submission can be done online via the corporate requirements management tool.

B.5.3.2. *Schedule control plan*

Once the master schedule is baselined, progress of the project will be tracked on a weekly basis via status meetings of individual teams. The project manager will chair the status meeting (that all teams are participating) for the entire project. SPI will be calculated weekly based on progress of all teams. SPI will be controlled within 5% of 1.0.

B.5.3.3. *Budget control plan*

Similar to schedule control, cost of the project will be tracked weekly. The actual spending will be compared against the baseline. CPI will be calculated weekly and controlled within 5% of 1.0.

B.5.3.4. *Quality control plan*

The defect rate must be above 5 Sigma in terms of Lines of Code (less than 1 defect per 100,000 lines of code). User trial tests must be stress-free for 1 hour of use. In order to achieve these quality goals, formal reviews will be conducted for all artifacts and metrics will be kept and reported weekly as part of the status meeting. The quality team is responsible for collection of appropriate metrics for derivation of the Sigma rating. The system test team is responsible for conducting and reporting the user trial tests.

B.5.3.5. *Reporting plan*

Status reports on the project will be published weekly based on the above control plans and posted on the project library. Each team will post its own report and the project manager will post the report for the entire project.

B.5.3.6. *Metrics collection plan*

The quality team is responsible for collection of appropriate metrics. All teams are expected to cooperate with the quality team for this purpose (e.g., provide line counts for the product). The quality metrics will be detailed in the quality assurance plan and posted on the project library. The quality team will provide the weekly metrics for individual teams and the entire project for incorporation into their respective reports.

B.5.4. Risk management plan

Initial risks are derived from project assumptions and constraints. Additional risks will be identified by brainstorming at the beginning of the project and further identified by all team members as the project progresses. Each team will actively tackle its top three risks and the project manager will oversee the top five risks for the entire project. Time will be allocated in weekly team meetings and weekly project meeting to discuss progress in mitigating risks. Risks will be posted and updated weekly on the project library.

B.5.5. Closeout plan

Individual phases will be closed out at the team and project levels. At the end of the project, the entire project will be closed by contract closure with the outsource/offshore entity; sign-off by DND, CNA, and other stakeholders; and administrative closure of the project by archiving all artifacts in the project library. A "lessons learned" session will be conducted too.

B.6. Technical Process Plans

B.6.1. Process model

The traditional waterfall model will be used for software development. However, there will be multiple development iterations for both the control and database modules, and a prototype for the product will be available one year after the project starts.

B.6.2. Methods, tools and techniques

It will be up to the individual development teams to identify additional software development tools and test tools to complement the basic Windows 7 32-bit and 64-bit versions; Linux 3.5; Snow Leopard 9.1; and Java version 7.1. Software budget has been allocated for the purchase of these tools.

B.6.3. Infrastructure plan

Corporate infrastructure currently in place will be sufficient to serve the duration of this project. No additional requirement is identified at this moment.

B.6.4. Product acceptance plan

The system test team will set up the test lab which will also be used to demonstrate the prototype and final product to the stakeholders and (selected) end users. Upon completion of our own system test, stakeholders and (selected) end users will be invited to conduct their acceptance test on the prototype and final product for sign-off.

B.7. Supporting Process Plans

B.7.1. Configuration management plan

The existing corporate configuration management policy (see company intranet web page) applies to and is adequate for this project. All artifacts (e.g., documents, code, lab equipment) are under configuration management.

B.7.2. Verification and validation plan

The system test team is responsible for verification and validation of the prototype and final product. The team is responsible for derivation of the test plan, test scenarios (i.e., environment setup) and test cases. Test results will be captured and appropriate metrics are reported during the test phase.

B.7.3. Documentation plan

All internal documents for the project (e.g., PMP, design document, test cases) are posted on the project library. External documents (e.g., requirements document, user manuals) are delivered per contract agreement at the beginning of the project, or as part of the prototype and final product to the stakeholders and end users for sign-off. These documents are also posted on the project library.

B.7.4. Quality assurance plan

The quality team is responsible for this plan. The plan is aligned with the corporate quality goals and amended with project specific goals. The defect rate must be above 5 Sigma in terms of Lines of Code (less than 1 defect per 100,000 lines of code). User trial tests must be stress-free for 1 hour of use. The quality team is going to derive appropriate quality metrics for collection during project execution in order to fulfill the quality goals.

B.7.5. Reviews and audits

All artifacts for the project are subject to formal technical reviews. The corporate formal review process will be followed.

Audits on the development process are to be conducted by the quality team during the course of running the project. Quality metrics will be captured based on audit results.

B.7.6. Problem resolution plan

Problems identified during formal reviews will be closed by the individuals responsible for the artifacts. The turnaround time for closing these problems is in accordance with the project schedule.

Problems identified by the test team during system test are to be fixed by the appropriate development team. The turnaround time for closing these problems is normally within one week, or sooner depending on the severity of the problems.

Problems identified by the stakeholders and end users after product deployment are to be fixed by the appropriate development team. The turnaround time for closing these problems are normally within one month (software will be formally released once a month), or sooner depending on the severity of the problems. For critical problems, fixes can be provided as workaround (i.e., patches) instead of a formal release.

B.7.7. Subcontractor management plan

The outsource/offshore entity is to be managed by the project manager and team leaders, together with the staff overseeing database module development. Focal points for regular contact and focal points for escalation must be identified on both sides. Weekly status meeting (just like Vacux internal meetings) will be held to obtain formal update from the outsource/offshore entity. Appropriate project metrics (e.g., SPI, CPI) and quality metrics are generated.

B.7.8. Process improvement plan

The quality team will help identify the top 3 process improvement initiatives for the development teams to work on. These will be captured in the teams' schedules in order to track the progress to completion. Brainstorming amongst development teams will be conducted to solicit initial ideas. The quality team will then facilitate the prioritization with a scoring algorithm and pick the top 3 to work on.

B.8. Additional Plans

B.8.1. Safety

N/A.

B.8.2. Privacy

Confidentiality of database information due to outsource/offshore development of the database module will be a concern. This will be left as an open issue for the time being and a work group will be formed between Vacux and the outsource/offshore entity to iron out the problems.

B.8.3. Security

Database security will be a concern due to outsource/offshore development of the database module. This will be left as an open issue for the time being and a work group will be formed between Vacux and the outsource/offshore entity to iron out the problems.

B.8.4. Installation/rollout

Deployment across NA will be done by electronic software distribution. This is a standard process of Vacux for all software deployment and will be tested as part of the system test.

B.8.5. Maintenance

Ongoing maintenance of the software modules will be the responsibilities of Vacux and the outsource/offshore entity. Two support persons (one on each side) will be staffed for this purpose.

B.8.6. Additional subplans

None.

Exercise

The above IEEE PMP Sample for SPENTFUEL is incomplete. Some key points and details were left out on purpose. Identify these deficiencies!

APPENDIX C

10% Table

N	F/P	P/F	A/F	F/A	A/F	F/A
1	1.1	0.909091	1	1	1.1	0.909091
2	1.21	0.826446	0.47619	2.1	0.57619	1.735537
3	1.331	0.751315	0.302115	3.31	0.402115	2.486852
4	1.4641	0.683013	0.215471	4.641	0.315471	3.169865
5	1.61051	0.620921	0.163797	6.1051	0.263797	3.790787
6	1.771561	0.564474	0.129607	7.71561	0.229607	4.355261
7	1.948717	0.513158	0.105405	9.487171	0.205405	4.868419
8	2.143589	0.466507	0.087444	11.43589	0.187444	5.334926
9	2.357948	0.424098	0.073641	13.57948	0.173641	5.759024
10	2.593742	0.385543	0.062745	15.93742	0.162745	6.144567
11	2.853117	0.350494	0.053963	18.53117	0.153963	6.495061
12	3.138428	0.318631	0.046763	21.38428	0.146763	6.813692
13	3.452271	0.289664	0.040779	24.52271	0.140779	7.103356
14	3.797498	0.263331	0.035746	27.97498	0.135746	7.366687
15	4.177248	0.239392	0.031474	31.77248	0.131474	7.60608
16	4.594973	0.217629	0.027817	35.94973	0.127817	7.823709
17	5.05447	0.197845	0.024664	40.5447	0.124664	8.021553
I8	5.559917	0.179859	0.02193	45.59917	0.12193	8.201412
19	6.115909	0.163508	0.019547	51.15909	0.119547	8.36492
20	6.7275	0.148644	0.01746	57.275	0.11746	8.513564
21	7.40025	0.135131	0.015624	64.0025	0.115624	8.648694
22	8.140275	0.122846	0.014005	71.40275	0.114005	8.77154
23	8.954302	0.111678	0.012572	79.54302	0.112572	8.883218

(Continued)

(*Continued*)

N	F/P	P/F	A/F	F/A	A/F	F/A
24	9.849733	0.101526	0.0113	88.49733	0.1113	8.984744
25	10.83471	0.092296	0.010168	98.34706	0.110168	9.07704
26	11.91818	0.083905	0.009159	109.1818	0.109159	9.160945
27	13.10999	0.076278	0.008258	121.0999	0.108258	9.237223
28	14.42099	0.069343	0.007451	134.2099	0.107451	9.306567
29	15.86309	0.063039	0.006728	148.6309	0.106728	9.369606
30	17.4494	0.057309	0.006079	164.494	0.106079	9.426914
31	19.19434	0.052099	0.005496	181.9434	0.105496	9.479013
32	21.11378	0.047362	0.004972	201.1378	0.104972	9.526376
33	23.22515	0.043057	0.004499	222.2515	0.104499	9.569432
34	25.54767	0.039143	0.004074	245.4767	0.104074	9.608575
35	28.10244	0.035584	0.00369	271.0244	0.10369	9.644159
36	30.91268	0.032349	0.003343	299.1268	0.103343	9.676508
37	34.00395	0.029408	0.00303	330.0395	0.10303	9.705917
38	37.40434	0.026735	0.002747	364.0434	0.102747	9.732651
39	41.14478	0.024304	0.002491	401.4478	0.102491	9.756956
40	45.25926	0.022095	0.002259	442.5926	0.102259	9.779051
41	49.78518	0.020086	0.00205	487.8518	0.10205	9.799137
42	54.7637	0.01826	0.00186	537.637	0.10186	9.817397
43	60.24007	0.0166	0.001688	592.4007	0.101688	9.833998
44	66.26408	0.015091	0.001532	652.6408	0.101532	9.849089
45	72.89048	0.013719	0.001391	718.9048	0.101391	9.862808
46	80.17953	0.012472	0.001263	791.7953	0.101263	9.87528
47	88.19749	0.011338	0.001147	871.9749	0.101147	9.886618
48	97.01723	0.010307	0.001041	960.1723	0.101041	9.896926
49	106.719	0.00937	0.000946	1057.19	0.100946	9.906296
50	117.3909	0.008519	0.000859	1163.909	0.100859	9.914814
52	142.0429	0.00704	0.000709	1410.429	0.100709	9.929599
53	156.2472	0.0064	0.000644	1552.472	0.100644	9.935999
54	171.8719	0.005818	0.000585	1708.719	0.100585	9.941817

(*Continued*)

(Continued)

N	F/P	P/F	A/F	F/A	A/F	F/A
55	189.0591	0.005289	0.000532	1880.591	0.100532	9.947106
56	207.9651	0.004809	0.000483	2069.651	0.100483	9.951915
57	228.7616	0.004371	0.000439	2277.616	0.100439	9.956286
58	251.6377	0.003974	0.000399	2506.377	0.100399	9.96026
59	276.8015	0.003613	0.000363	2758.015	0.100363	9.963873
60	304.4816	0.003284	0.00033	3034.816	0.10033	9.967157
61	334.9298	0.002986	0.000299	3339.298	0.100299	9.970143
62	368.4228	0.002714	0.000272	3674.228	0.100272	9.972857
63	405.2651	0.002468	0.000247	4042.651	0.100247	9.975325
64	445.7916	0.002243	0.000225	4447.916	0.100225	9.977568
65	490.3707	0.002039	0.000204	4893.707	0.100204	9.979607
66	539.4078	0.001854	0.000186	5384.078	0.100186	9.981461
67	593.3486	0.001685	0.000169	5923.486	0.100169	9.983147
68	652.6834	0.001532	0.000153	6516.834	0.100153	9.984679
69	717.9518	0.001393	0.000139	7169.518	0.100139	9.986071
70	789.747	0.001266	0.000127	7887.47	0.100127	9.987338
71	868.7217	0.001151	0.000115	8677.217	0.100115	9.988489
72	955.5938	0.001046	0.000105	9545.938	0.100105	9.989535
73	1051.153	0.000951	9.52E-05	10501.53	0.100095	9.990487
74	1156.269	0.000865	8.66E-05	11552.69	0.100087	9.991351
75	1271.895	0.000786	7.87E-05	12708.95	0.100079	9.992138
76	1399.085	0.000715	7.15E-05	13980.85	0.100072	9.992852
77	1538.993	0.00065	6.5E-05	15379.93	0.100065	9.993502
78	1692.893	0.000591	5.91E-05	16918.93	0.100059	9.994093
79	1862.182	0.000537	5.37E-05	18611.82	0.100054	9.99463
80	2048.4	0.000488	4.88E-05	20474	0.100049	9.995118
81	2253.24	0.000444	4.44E-05	22522.4	0.100044	9.995562
82	2478.564	0.000403	4.04E-05	24775.64	0.10004	9.995965
83	2726.421	0.000367	3.67E-05	27254.21	0.100037	9.996332
84	2999.063	0.000333	3.34E-05	29980.63	0.100033	9.996666

(Continued)

(Continued)

N	F/P	P/F	A/F	F/A	A/F	F/A
85	3298.969	0.000303	3.03E-05	32979.69	0.10003	9.996969
86	3628.866	0.000276	2.76E-05	36278.66	0.100028	9.997244
87	3991.753	0.000251	2.51E-05	39907.53	0.100025	9.997495
88	4390.928	0.000228	2.28E-05	43899.28	0.100023	9.997723
89	4830.021	0.000207	2.07E-05	48290.21	0.100021	9.99793
90	5313.023	0.000188	1.88E-05	53120.23	0.100019	9.998118
91	5844.325	0.000171	1.71E-05	58433.25	0.100017	9.998289
92	6428.757	0.000156	1.56E-05	64277.57	0.100016	9.998444
93	7071.633	0.000141	1.41E-05	70706.33	0.100014	9.998586
94	7778.796	0.000129	1.29E-05	77777.96	0.100013	9.998714
95	8556.676	0.000117	1.17E-05	85556.76	0.100012	9.998831
96	9412.344	0.000106	1.06E-05	94113.44	0.100011	9.998938
97	10353.58	9.66E-05	9.66E-06	103525.8	0.10001	9.999034
98	11388.94	8.78E-05	8.78E-06	113879.4	0.100009	9.999122
99	12527.83	7.98E-05	7.98E-06	125268.3	0.100008	9.999202
100	13780.61	7.26E-05	7.26E-06	137796.1	0.100007	9.999274
101	15158.67	6.6E-05	6.6E-06	151576.7	0.100007	9.99934
102	16674.54	6E-05	6E-06	166735.4	0.100006	9.9994
103	18342	5.45E-05	5.45E-06	183410	0.100005	9.999455
104	20176.19	4.96E-05	4.96E-06	201751.9	0.100005	9.999504
105	22193.81	4.51E-05	4.51E-06	221928.1	0.100005	9.999549
106	24413.2	4.1E-05	4.1E-06	244122	0.100004	9.99959
107	26854.51	3.72E-05	3.72E-06	268535.1	0.100004	9.999628
108	29539.97	3.39E-05	3.39E-06	295389.7	0.100003	9.999661
109	32493.96	3.08E-05	3.08E-06	324929.6	0.100003	9.999692
110	35743.36	2.8E-05	2.8E-06	357423.6	0.100003	9.99972
111	39317.7	2.54E-05	2.54E-06	393167	0.100003	9.999746
112	43249.46	2.31E-05	2.31E-06	432484.6	0.100002	9.999769
113	47574.41	2.1E-05	2.1E-06	475734.1	0.100002	9.99979
114	52331.85	1.91E-05	1.91E-06	523308.5	0.100002	9.999809

(Continued)

(Continued)

N	F/P	P/F	A/F	F/A	A/F	F/A
115	57565.04	1.74E-05	1.74E-06	575640.4	0.100002	9.999826
116	63321.54	1.58E-05	1.58E-06	633205.4	0.100002	9.999842
117	69653.7	1.44E-05	1.44E-06	696527	0.100001	9.999856
118	76619.07	1.31E-05	1.31E-06	766180.7	0.100001	9.999869
119	84280.97	1.19E-05	1.19E-06	842799.7	0.100001	9.999881
120	92709.07	1.08E-05	1.08E-06	927080.7	0.100001	9.999892

APPENDIX D

SEI-CMM Ladder

Based in Pittsburgh, Pennsylvania, USA, the Software Engineering Institute (SEI) published the Capability Maturity Model (CMM) to measure the maturity of software development. SEI-CMM was based on the process maturity framework created by Humphrey [1]. It is widely used by software organizations world-wide for understanding, evaluating and improving their process capability and maturity. The CMM was published as a book by Paulk *et al.* [2]. There are five levels of maturity, each with defined process areas and practices, as shown in Figure D.

The assessment methodology relies on a process maturity questionnaire of 124 items with Yes/No answers. About 90% of key questions and 80% of all questions for a level must be answered Yes in order for an organization to qualify that level. One must fully qualify a level before being considered for the next level. As shown in Figure D, all organizations, by default, start at Level 1: Initial. Subsequent levels each has a set of defined process areas and practices to tackle. We do not intend to explore these in details now. However, it usually takes at least two years for an organization to move up one step of the SEI-CMM ladder.

Executive commitment is definitely needed to achieve such goal, as this is going to bear certain cost for the organization, with the payback being improvement in quality and productivity. One has to carefully build a business case with benefit-cost analysis to justify the work.

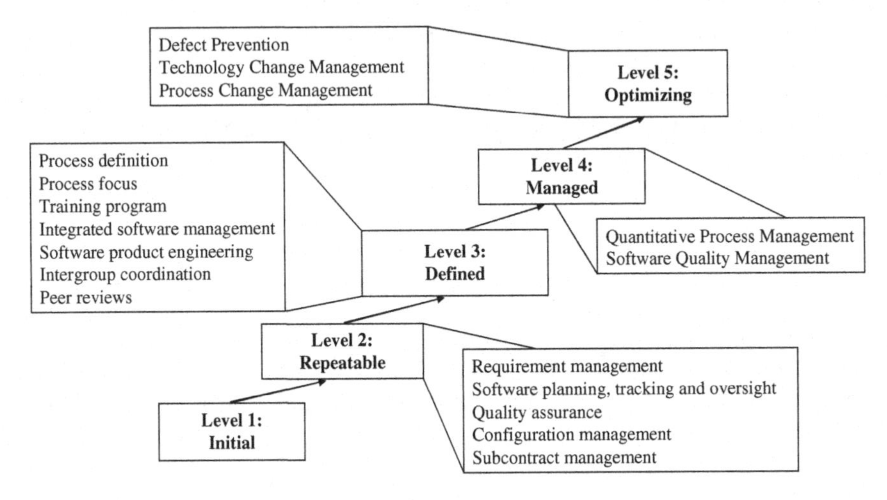

Figure D. SEI-CMM ladder.

In brief, the five SEI-CMM levels are:

1. Initial — when things are chaotic and *ad hoc* without any documented or repeatable process.
2. Repeatable — steps can be repeated as per documented process in building software.
3. Defined — the process is defined and confirmed as a standard business process (as is).
4. Managed — the process is quantitatively measured based on agreed-upon metrics.
5. Optimized — the process is being continuously improved and optimized through adoption of new technology and process.

References

1. Humphrey, W. *Managing the Software Process*. Addison Wesley, 1989.
2. Paulk, M. C., Weber, C. V., Curtis, B. and Chrissis, M. B. *The Capability Maturity Model: Guidelines for Improving the Software Process.* Boston: Addison Wesley, 1995.

APPENDIX E

Survey Questions

Background: X has been helping you a lot throughout your lifespan of 20 odd years. You made a commitment to X to discuss matters (deemed unimportant to you) over the phone for 15–30 minutes on a certain day.

For each of the following 10 questions, answer with ratings: very likely (5), somewhat likely (4), neutral (3), somewhat unlikely (2), or very unlikely (1).

Q1. During your hectic and adverse moments (assignments due, term test approaching, feeling lonely and depressed), how likely is it for you to forget about the commitment?

Q2. Suppose you did not forget about the commitment, yet you knew you could not make it. How likely is it for you to notify X prior to or on the committed date?

Q3. Suppose X was quite understanding and agreed with you to postpone the discussion to a later date. You were still caught in a spiral of more hectic moments and adverse events. How likely is it for you to meet this postponed commitment (still deemed unimportant to you)?

Q4. Suppose you knew you could not make it and sent an email to X requesting further delay. Unfortunately, due to timing issues in communication (email delay, different time zones), no confirmation or acknowledgement was received from X. How likely is it for you to notify X via other means of communication (e.g., phone call) prior to the scheduled date and time?

Q5. At the same time, you were caught in an after-school event in the evening. All students attended a presentation by a potential employer. The presentation started more than an hour late and was going to overrun till almost 11 p.m. With your current study load, you needed the time this evening to finish some assignments due the next day. How likely is it for you to walk out of this presentation?

Q6. Considering your commitment to X, suppose your email requesting further delay was never acknowledged and you did not attempt to notify X via other means of communication. During the after-school presentation (started late and overran), it was also time for you to have the 15–30 minutes phone discussion with X. How likely is it for you to get out of the presentation briefly and notify X of your situation?

Q7. Suppose you stayed at the presentation till the end and never made an effort to call X. By the time you arrived home, still having to finish the assignments due the next day, still have not had dinner, etc., you saw an angry email from X after waiting for your call and blaming you for repeatedly missing your commitments. How likely is it for you to feel being at fault?

Q8. How likely is it for you to apologize and commit to a date/time that you would try your very best not to miss again?

Q9. How likely is it for you to sort through your grievances (study load, depression, etc.) and use these as your excuses of missing your commitments?

Q10. Suppose you poured your heart out, only to be met with more furious responses from X, saying that you were incapable and irresponsible. How likely is it for you to de-commit from the 15–30 minutes discussion with X, seeing that you have too many important and higher priority items to handle?

While the tabulated survey results have been given in Chapter 14, we also present some written comments here. On one hand, we note the following:

"Person X should also realize people have some other things to do and should not put too much pressure on anyone."

"X seems like parents in this situation. I feel that parents may be understanding and their demands and expectations are not high."

"I no longer care to discuss the matter."

On the other hand, we note the following:

"I try to meet all commitments and hope to do so in the future."

"The issue here is very poor time management skills, lack of foresight and pro-activity; causing these scheduling problems in the first place. There is no reason for a 15-30 minute phone appointment to be missed."

"I am a reasonably organized person and have never had so much work that I could not find time, and plan around a 15–30 minutes call. I meet my commitments."

APPENDIX F

Offshoring Benefit-Cost Analysis

F.1. Introduction

Over the past decades, companies in various fields, especially IT [1], are jumping on the bandwagon of offshoring. Is it cost beneficial after all? How can we estimate the benefit and cost of such adventures?

This appendix attempts to answer some of the above queries by looking at the benefit and cost estimation for software offshoring, based on practical experience in the industry, in two ways: Intra-Firm Offshoring [2] to an offshore affiliated entity that the company itself owns, and Third-Party Offshoring to an offshore unaffiliated company.

We derive and fine-tune the formulae for benefit-cost analysis. The formulae are presented in a generic manner as no tangible data points can be disclosed due to confidentiality.

F.2. Intra-Firm Offshoring

For simplicity, we assume the company originates in North America, but the same can be applied to European and other countries. Typically, North American companies use a generic cost per staff-year, say C_A = US$150,000. This figure is derived from the average salary (including benefits such as pension, medical, etc.) of all staff (from executives, managers to the most junior staff), plus fixed cost like desktop, laptop, hardware and software, office space, travel, expense, etc. For the offshore entity that the company itself owns, the generic cost per staff-year is of course much lower, say C_o = US$20,000. Depending on the exact location (e.g., India [3], China [3], Poland, the Czech Republic), C_o varies, but it is always much lower than the corresponding North American company.

The offshoring process involves three steps: transition of knowledge, joint development for the first project and offshore development of subsequent projects.

We estimate the cost of transition to be:

$$(A \times C_A + I \times C_o) \times m/12 \quad (1)$$

assuming *A* people from North America and *I* people from offshore are involved in the process of knowledge transition, and training for m months.

We estimate the saving of the first project to be:

$$((J - B) \times C_A - J \times C_o) \times n/12 \quad (2)$$

assuming joint development (hands-on training, paired programming, etc.) involving *B* people from North America and *J* people from offshore for a *n*-month project. The net may be positive or negative in this case.

Note that the practice of a round of joint development is entirely optional. If executives deem the knowledge transition to be successful, the offshoring entity may be asked to step up and undertake development immediately.

The true saving will be observed for subsequent projects:

$$(C_i \times C_A - K_i \times C_o) \times p_i/12 \quad (3)$$

assuming saving of C_i people from North America and cost of K_i people from offshore for p_i months in the *i*th project.

C_i should be the same as K_i, but at times, executives would allow $K_i > C_i$, due to low cost of the offshore entity. In addition,

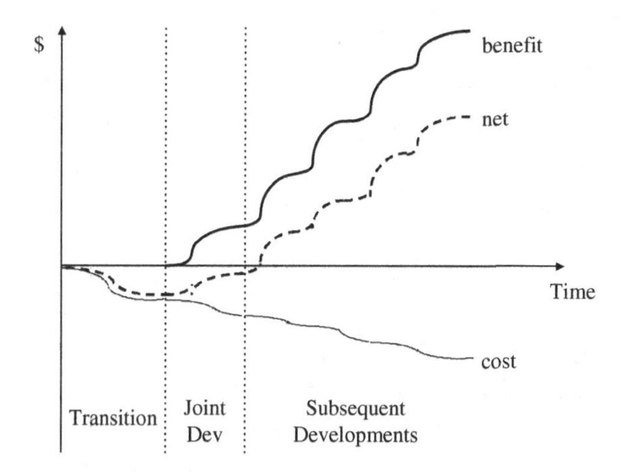

Figure F.1. Benefit-cost analysis of intra-firm offshoring.

executives are in favour of and hoping to shorten the development time by throwing more people into the project.

The benefit-cost analysis follows general S-curves [4] in Figure F.1. However, the exact shape and form of each S-curve depend on the project parameters (duration, effort, staffing, etc.) or the values of A, B, C_i, I, J, K_i, m, n, p_i, C_A, and C_o.

The formula for benefit-cost analysis (positive for saving, negative for cost) is given by combining (1), (2) and (3):

$$\begin{aligned}&\textstyle\sum_{i=2,..\infty}(C_i \times C_A - K_i \times C_o) \times p_i/12 + \\ &((J - B) \times C_A - J \times C_o) \times n/12 - \\ &(A \times C_A + I \times C_o) \times m/12\end{aligned} \tag{4}$$

The expectation is that the initial cost will be offset by savings accrued in subsequent projects.

F.3. Third Party Offshoring

This case is very straightforward, as the cost will simply be quoted by the third-party company, and invoiced to the originating company when the project is over and exit criteria are satisfied. The benefit will be the staff saving in the originating company.

Depending on the actual situation, the offshoring process can be simplified. Knowledge transition and training may still be needed, but joint development for the first project is usually skipped, as the two companies are different.

The cost of transition is:

$$A \times C_A \times m/12 + X \tag{5}$$

assuming A people from North America for m-month transition and \$X is charged by the offshore company.

Saving for each subsequent project will be:

$$B_i \times C_A \times n_i/12 - Y_i \tag{6}$$

assuming saving of B_i people from North America for n_i months and \$$Y_i$ is charged by the offshore company for the ith project.

The formula for benefit-cost analysis (positive for saving, negative for cost) is given by combining (5) and (6):

$$\Sigma_{i=1..\infty}(B_i \times C_A \times n_i/12 - Y_i) - (A \times C_A \times m/12 + X) \tag{7}$$

The benefit-cost analysis follows the general trend in Figure F.2. The benefit follows general S-curves. The cost follows step functions as lump sums charged by the offshore company. The exact shape and form of each S-curve and step function depend on A, B_i, m, n_i, X, Y_i, and C_A.

Again, the initial cost will be offset by savings accrued in subsequent projects.

F.4. Discussion

During the planning phase, based on analysis of the planning items listed in Chapter 12, the variables identified in Sections F.2 and F.3 (i.e., A, B, C, I, J, K, m, n, p, X, Y, C_A, C_o, etc.) and Equations (1)–(7) can be refined to give a more accurate benefit-cost estimation.

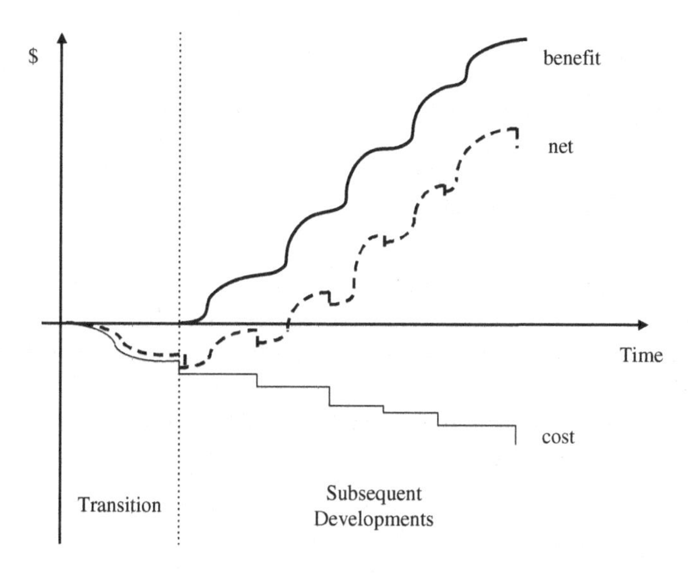

Figure F.2. Benefit-cost analysis of third-party offshoring.

During the execution phase, the actuals will be captured for Equations (1)–(7) and for derivation of the values of all variables in the equations. A set of automated tools should be deployed to oversee all aspects of the operation (i.e., transition, actual development) and to measure tangible benefit-cost of offshoring.

Measurement is difficult, as there is a tendency not to measure [5]. What criteria will be used by the originating entity to stop the transition or take back subsequent developments due to non-performance or unsatisfactory performance of the offshore entity? From a benefit-cost analysis point of view, the determining criteria would be the net resulting from subtracting the benefit from cost for any project.

During the closing phase, we put an orderly end to the transition and development, regardless of smooth transition, successful development or abrupt termination. The expense of the transition and tangible saving in development after the transition are captured and reported. The actuals for Equations (1)–(7) can be finalized.

References

1. National Academy of Engineering. *The Offshoring of Engineering*. The National Academies Press, 2008.
2. Contractor, F. J., Kumar, V., Kundu, S. K. and Pedersen, T. *Global Outsourcing and Offshoring*. Cambridge, MA: Cambridge University Press, 2011.
3. Meredith, R. *The Elephant and the Dragon: The Rise of India and China and What It Means for All of Us*. New York: W.W. Norton, 2007.
4. Putnam, L. H and Myers, W. *Measures for Excellence*. Prentice Hall, 1992.
5. Preston, R. Outsourcing Customers Send Some Mixed Signals. *Informationweek*, Feb 2007.

APPENDIX G

Project Management Professional Certification

This appendix covers the education and experience requirements, certification examination preparation, examination hints and tips, and ongoing professional development requirements for Project Management Professional (PMP) certification. The information is current as of the time this book was written. Refer to the Project Management Institute (PMI) website.

G.1. Educational and Experience Requirements

For candidates with university degrees at the time of application, the following experience requirements are needed:

- at least 3 years of unique (i.e., non-overlapping) project management experience within the six-year period prior to application,
- at least 4500 hours of project management experience within the five process groups (i.e., initiating, planning, executing, monitoring and controlling, and closing), and
- 35 contact hours of project management education.

For candidates with high school diplomas or secondary school credentials, the following experience requirements are needed:

- at least 5 years of unique (i.e., non-overlapping) project management experience within the eight-year period prior to application,

- at least 7500 hours of project management experience within the five process groups (i.e. initiating, planning, executing, monitoring and controlling, and closing), and
- 35 contact hours of project management education.

Each candidate has to submit an application, somewhat like a resume, preferably on-line (even though paper submission is also accepted). The candidate has to provide details of each project she manages, broken down into hours spent in the five process groups. It is understood that there may not be coverage in certain process group(s) for certain project(s). For example, one who gets transferred into management of a project already under execution will have zero initiating and planning hours for the project.

Note also that one must be able to reasonably account for the hours spent on a project. For example, if one works 40 hours a week, yet 50% on managing a project (50% on non-management activities), with productivity of 80%, then there will be 16 hours per week spent on the project. If the project lasts 10 weeks, then 160 hours (plus overtime, of course) will be expended across the five process groups.

G.2. Eligibility for PMP Certification Examination

Once the application is approved, an eligibility letter will be issued to confirm a candidate's eligibility to take the examination. The eligibility period is six months, extensible for an additional six months. A candidate has to book the time to write the examination at an examination center from a location of his choice (e.g., close to work, close to home, etc.).

G.3. Certificate Examination

The PMP examination is computer based. All supplies (paper, pencil, etc.) as well as a five-function calculator will be provided. There will be a 15-minute online tutorial in order to familiarize with the computer. The examination will last for 4 hours and there will be 200 four-option multiple choice questions. One must obtain a score of 137 to pass. The result is presented to you at the end of the examination with the exact score for each process/area, but the bottom-line is: you either pass or fail.

Table G is a typical breakdown of the examination questions [1].

Table G. Breakdown of PMP examination questions.

Process/Area	Initiating	Planning	Executing	Monitoring and controlling	Closing	Professional responsibility
Number of questions	17	47	47	46	14	29
Percentage (%)	8.5	23.5	23.5	23.0	7.0	14.5

G.4. Hints and Tips for the PMP Examination

1. Know your formulae

This is the most important point. Write down the formulae (e.g., EV, CPI, SPI) while your brain is clear. It is suggested that you "core dump" everything to a piece of paper at the start of the examination, or even smarter, during the 15 minutes online tutorial session.

2. Read each question twice

You have a little more than 1 minute per question. Read each question twice to make sure that you fully understand it, and read all four choices so that you do not get too excited by a correct (but not the most correct) answer.

3. Mark the doubtful questions

Some questions are factual while others are hypothetical. Doubtful questions can be guess-answered, marked, and revisited at a later time.

4. Select the best answer
   Eliminate incorrect answers to improve your probability of selecting the correct one. For some questions, all four choices are correct. Pick the most correct answer and be familiar with PMI's perspective, not the perspective you acquired from life.
5. Remember the areas that PMI emphasizes.
   - Planning is the most important process group.
   - Team must be involved early and in all major decisions.
   - Lessons learned and historical records are essential inputs.

G.5. Professional Development Requirements

One has to renew her certification over a 3-year period by attaining 60 Professional Development Units (PDUs). PDUs are earned by reporting qualified activities on-line. Some examples (directly related to project management) are training (offering or attending), publishing papers, writing books, giving presentations, mentoring other professionals, organizing PMI activities, etc.

One must also sign and adhere to the PMI PMP Code of Professional Conduct.

G.6. Other PMI Examinations

There are other PMI examinations for the following designations:

- Certified Associate in Project Management (CAPM)
- Program Management Professionals (PgMP)
- PMI Agile Certified Practitioner (PMI-ACP)
- PMI Risk Management Professional (PMI-RMP)
- PMI Scheduling Professional (PMI-SP)

Refer to the PMI website for details.

References

1. Available at: http://www.pmi.org/certification/project-management-professional-pmp/pmp-exam-prep.aspx.

APPENDIX H

PRINCE, APM, MSP Roadmaps

This appendix covers the certification roadmaps for several project management standards. The information is current as of the time this book was written.

H.1. PRINCE Roadmap

Project IN Controlled Environment (PRINCE) is the *de facto* standard used by the UK government. It was established in 1989 by the Central Computer and Telecommunications Agency (CCTS), now the Office of Government Commerce (OGC). PRINCE 2 was published in 1996.

PRINCE is process-based method for effective project management and is focused on business justification. It follows a product-based planning approach with emphasis on dividing projects into manageable and controllable stages. The organization structure for the project management team is well-defined. PRINCE provides overall flexibility to be applied at a level appropriate to a project.

- PRINCE 2 Foundation

At a cost of £190, this is a 1-hour, closed book examination to test a candidate's understanding of PRINCE 2. You need 38/75 multiple choices to pass.

- PRINCE 2 Practitioner
  At a cost of £350, this is a 3-hour, open book examination to test a candidate's application of PRINCE 2. You need 180 marks out of nine 40-mark essay questions to pass.

H.2. APM Roadmap

The following describes the Association of Project Management (APM) four-tier roadmap.

- APM Introductory Certificate
  This is a 1-hour, 60 multiple choices examination.
- APMP
  This is a 3-hour essay examination.
- APM Practitioner
  One has to go into an assessment center for two days, participate in individual and group work, and go through interviews.
- Certified Project Manager (CPM)
  One has to submit an application, complete a project report, and go through a panel interview.

H.3. MSP Roadmap

The Managing Successful Program (MSP) roadmap is similar to the PRINCE 2 roadmap with a two-tier certification: MSP Foundation and MSP Practitioner.

Index

Words of Wisdom for Engineering Project Managers

We offer a prize to the reader who can identify the famous people who offered these words of wisdom.

"There is no elevator to success: one has to take the stairs."

"Every success is built on the ability to do better than just good enough."

"Whenever an individual or a business decides that success has been attained, then progress stops."

"Difficulties are meant to make us better, not bitter."

"The pessimist sees the difficulty in every opportunity, the optimist, the opportunity in every difficulty."

"Cause change and lead, accept change and survive, resist change and die."

"The most effective way to cope with change is to help create it."

"A goal without a plan is just a wish."

"Experience is not what happens to you; it is what you do with what happens to you."

"If you do the little jobs well, the big ones will tend to take care of themselves."

"Your mind is like a parachute, it works best when it is opened."

Printed in the United States
By Bookmasters